Fish Byne Yola

The Jew Who Saved Galileo

Copyright 2016
ISBN-13: 978-1540309327
ISBN-10: 1540309320

Cast of Characters

Salvatore Isaac Fishbeingnola – Jew passing as Catholic, Medical Doctor, One Man Greek Chorus, Self-Appointed Protector of Galileo

Galileo Galilei – Tinkerer, Natural Philosopher, Possible Heretic, Illegitimate Parent, Smart Ass, Putative Creator of the Scientific Method

Roberto Bellarmino – Jesuit, Bishop, Cardinal, Inquisitor, Doctor of the Church, Canonized by Pope Pius XI, buried at the Church of Saint Ignatius

Maffeo Bellarmino – Roberto's brother, not Jesuit, not Bishop, not Cardinal, not Doctor of the Church, not canonized, reprobate, madman, burial place unknown

Father Francesco Montepulciano – acolyte to Roberto Bellarmino, sociopath

Bufana Giacometti – passing love interest for Galileo, seed of conflict

Vito Fishbeingnola – son to his widower father, Salvatore

Anna Bolen – hot Swede, nurse, housekeeper, lover to Fishbeingnola, surrogate mother to Vito

Federico Fangio (Fang) – friend to Vito, adopted son to Fishbeingnola, convert to Judaism

Marina Gamba –mother of Galileo's children

Amadeo – hired thug and snitch for Fishbeingnola

Guido (Il Venezorino) – friend to Amadeo, factotum to Fishbeingnola

Elie Moberger – Fang's father-in-law, aide to Fang's conversion to Judaism

Prologue

The World in 1564

In Asia, the Chinese were trying to absorb the Mongols whose last great emperor, Altan Khan, was holding out valiantly against the Chinese Ming Dynasty. The whole absorption thing was back on track after being hijacked, twelve years earlier, by the Shensi earthquake. The quake killed 830,000 Chinese, which complicated casualty reporting and therefore confused the Chinese generals.

All the way across Eurasia, Elizabeth I was still Queen of England. In 1564 she got her first horse drawn carriage. The Dutch had had such things for a few years already.

The Holy Roman Emperor, who thought he had only Holy Roman Catholics in his empire, found his world mingling with and surrounded by Protestants, arguing with Eastern Orthodox apostates, and constantly sparring with Muslims.

Speaking of which, Suleiman I was nearing the end of his conquests in the Middle East. It was as good as it ever got for the Turks.

European explorers were flopping around between discoveries, hardly Christianizing and/or enslaving enough New World Indians to mention, although French fur traders were up the Saint

Lawrence trying hard to do both.

A couple fellows of note were born in 1564: Galileo Galilei and William Shakespeare.

In the intervening months between their births, the subject of this book, Salvatore Isaac Fishbeingnola, was born.

***

OnMarch 15, 1564, Julian Calendar, Salvatore Fishbeingnola escaped down his mother's birth canal to join all sorts of other escapees in the city of Venice. Venezia itself was built by refugees, driven to a swamp by Attila. These refugees found that a swamp that was an effective barrier to the Huns' horses. On this cool spring morning in 1564 it was peopled by the descendants of those refugees plus escapees of another sort. Jews had come to Rome even before Christian times, and had been accepted and ostracized in various places and at various times all over Europe. Here in Venezia, the Capitol of the Republic of Venice, the supreme naval power of the Adriatic and hence controller of trade with the Middle East and beyond, they had been confined to a city within a city. In a city/state that had evolved from a cluster of shivering refugees to Queen of the Adriatic, the latest immigrants were appointed pawn brokers and de facto jailed in their own little town with their own little town square and their own little doorways. It was called Ghetto.

Not all Jews. Salvatore's father, Filippo Morris Fishbeingnola, was an Ashkenazi, with a few more European genes than average. He was tall, blonde, blue-eyed, and non-observant, since Judaism didn't make any more sense to him than did Catholicism. He sometimes wondered if a female ancestor had been raped by a Viking on a Sicilian beach, but he didn't wonder very much, since he had decided to use his European genes to make it easier to assimilate. He had named his son Salvatore, not for savior, but for saved, an ironic salute to Abraham, the tragicomic patriarch who was going to kill his son until the Lord told him it was only a joke. Filippo wasn't going to wait for holy jokes to save his son. He was going to bestow every advantage he could.

Salvatore's mother was more Semitic in appearance. With her nearly jet black hair and her deep brown almond-shaped eyes she could have been from any desert tribe, Arab or Hebrew. Her figure was trim rather than voluptuous, but if her eyes were peering from behind a veil in a Bedouin tent, they could draw any mortal man into dreams of rapture with the woman who owned them. Her given name was Giubba, but Filippo called her Gioia, for the joy she brought him every day. She would do the same for her only son, although she did have him circumcised.

Filippo thought a move from the ghetto would solve certain problems he associated with being a Jew, but he didn't want to move too far and draw attention to the fact that he was trying to assimilate.

He found an affordable place across the canal in
Zappa Court, just off Cannaregio Street. The family
could have been happy there, with Salvatore enrolled
in a strict Catholic school and perhaps destined for a
university, Bologna or Padua, where he could study
Medicine. Salvatore threw a shadow over that
happiness, ironically because of his interest in said
Medicine at an early age. He played doctor with a
little neighbor girl and she described his circumcised
penis to her mother. Filipo thought it a good time to
get out of town, so they moved to Pisa, where Galileo
was born and where he still lived, sans parents. Both
boys were eight, and they first met in school.

Filipo had made the family fortune in
brokering codfish, which he couldn't do in any
volume in Pisa, so he established a marble factory. It
wasn't a factory for producing construction or
statuary marble, but one for producing those little
glass spheres which were becoming very popular in
Italy. People had devised all sorts of games using
them, natural philosophers had devised little
experiments with them, and engineers saw all sorts of
possible projects, including putting the bell tower for
the Pisa Duomo on rollers. For this last project,
unfortunately, they only completed about half the
base before they lost interest. They were happy to
know the marbles worked well as rollers, and left the
job incomplete. No matter. To this day,
Fishbeingnola marbles are second only to Murano
glass sculptures as treasured collector's items from
the Italian glass artisan tradition. This clever
entrepreneurship by Filippo Fishbeingnola made him

even more wealthy than had codfish brokering, and he soon was a pillar of Pisa society and a patron of Pisa arts. About that time the Council for the Inquisition into Matters of the Faith published its latest little handbook on how to properly conduct an Inquisition. Filippo read through this piece, written in Latin, very slowly and very studiously. It made him decide that his son Salvatore should go to a very good, very strict, very deeply devout Catholic school. And this is where Galileo Galilei met Salvatore Isaac Fishbeingnola; the academy at the Camaldolese Monastery. All because of a circumcised penis.

Chapter One

Fra Domenico had sung the start of matins, one of the two canonical hours that were celebrated by group mass at the monastery. The other was vespers. The acolytes at the monastery were expected to perform prayers at the other hours, especially nones and complins, which Galileo did. At least, that's what he told Salvatore Fishbeingnola, who in turn affirmed that he did, also. Salvatore was lying. He suspected the Galileo was truthful. Galileo was devout enough. It's just that his mind wandered. This morning was a good example. Father Pietro had led two altar boys to the tabernacle to consecrate the host. It was the duty of one of the boys to light the holy incense candle that hung above the altar, and in doing so he had disturbed it considerably. The cord on which it hung was still swinging. Galileo had his finger to the pulse in his wrist.

"You doing that again?" Salvatore's voice was almost a stage whisper.

"Shhh," said Galileo.

"Are you?"

"Shhh. This is the most important part of the mass."

"Then why are you taking your pulse?"

"God is trying to tell me something."

"What? That you're alive?"

"If you shut up now, I'll tell you later."

***

In a corner of the stable Fra Basilio had allowed Galileo to set up a little tinkering space. It was hidden by a main brace of the stable and some rough boards. These things interfered with the line of sight but not with the smell. It was, after all, a stable. The space could hold two people and one experiment, providing the experiment didn't involve more than a couple of Galileo's 'toys', as Fra Basilio called the little gadgets Galileo used to flesh out his ideas. Only Fra Basilio and Galileo and Salvatore knew about it.

Three people and the toys and the smell made it uncomfortable for Salvatore, who didn't have Galileo's focus. Salvatore's head was filled with a young man's hormones, so he couldn't enjoy this great conspiracy that seemed entirely without sin. He fixed on a bent and rusted nail, closed his eyes and let the nail become a tanned, red-headed girl in a beckoning pose. He opened his eyes and spotted a knothole, dark and round. He closed his eyes and saw the head of a brunette walking away from him. He constructed the rest of her, making her buttocks sway gently as she walked. His mind called out to her. "You didn't tell me your name," it said. He let her go and opened his eyes. He saw Fra Basilio. Would he have to mention these thoughts in confession? He might as well, although it was a waste of time. The prurient thoughts were expected in the confessional. That's all the priests heard from boys in the monastery. Same old confessions. Same old penance.

Galileo on the other hand could subordinate the hormones to mathematics, at least when he was in his secret space. More importantly, neither Fra Basilio nor Galileo nor Salvatore could find anything in either the Ten Commandments or the church canons that prohibited a toy room in a stable. They all knew that if the abbot were told about it he would find the missing regulation. As long as only three of them knew, plus, of course, God, it was innocent. Galileo and Salvatore had to arrive at the toy room via separate routes, which made it seem all the more conspiratorial. In their minds, their hearts, their souls, though, they felt sure it was neither heretical nor apostatizing. Still, for Salvatore it would have been a bit of Hell if he didn't, for the sake of sanity, fantasize about what might be under the skirts of the peasant girls down the hill. He loved his friend Galileo, so he went to the smelly toy box and dreamed about one thing while Galileo thought about another. He felt it was a melding of poetry and mathematics.

"Are you sure you're comfortable with this?" He watched Galileo set up some sort of apparatus.

"With what?"

"Your sneaky little magic shows."

"I don't believe in magic. And it's not sneaky. It's just private. I want to keep these things to myself until I'm sure of them. And I can only get sure by testing them."

"And you're not worried about being a heretic or something?"

"I believe in our Lord Jesus and our Mother, the Church. But the world works differently than most people think. That's not Jesus' fault. It just works that way. What about you. Are you worried about being a heretic?"

"Me? I just want to get along. I want to go to medical school. I need to check all the boxes. Being a good Catholic is one of them, at least here in Italy, what with the Pope being here and all."

"And you believe in Jesus?"

"I say the words. I'm good at learning the words. And the friars and priests are good at teaching them. I sometimes think nobody listens to themselves when they say the words. I know I could say them and keep three tops spinning at the same time. Or say the Apostles' Creed and think about beautiful lady apostles at the same time."

"There were no lady apostles."

"See. That's why I worry about you. You are a true believer, but this doesn't look like you are. If we get caught playing with your toys you'll get into a lot of trouble. More than me, because everyone will know they're your toys."

"If you'll shut up and just watch, and follow me through all this, you see that these aren't toys."

"Whatever you say."

Galileo began stringing up bobs of various lengths. He wound up his timing devices, of which he had three. Actually, only two needed winding. The other was powered by a reservoir of water. He shook his head as he looked at them. "This is such a handicap. I'm sure I know what I'm doing, but it involves keeping time really well, and, except for the water clock, I don't trust these things. Those two spring powered things rely on a fixed force to turn the gears at set intervals, and I don't know if the force is all that fixed. Pasquale, the tinkerer, worked hard on these things, but it seems like you can count on the water clock to work the best. There's something about how nature works that makes me sure that Aristotle's Falling Down Force is more reliable than Pasquale's Spring Pushing Force. I have to think about that some more."

"What's to think? In this world things fall down and stop. That's how it works. That's natural motion you can count on. Fall down and stop.

Everything in this world falls down and stops, unless you keep pushing. And things up there," Salvatore indicated the heavens, "those things go around and around in circles, and never stop. God put them on crystal spheres and gave it all a push, way back at the start of things."

"Is that what you believe?"

"It's what they tell me here at the academy. Notice how well I repeat it. I'm a very good scholar. To be honest, and I can always be honest with you, I really don't give a little piece of a rat's little hole where the shit comes out. Actually, I'm more interested in the rat's little hole where the shit comes out. How that works. Then I go from rats to people, who have bigger holes where the shit comes out. But it probably works the same. If I understand things like that, I can be a doctor."

"See, that's why I like to do tests like this," Galileo said. "Even if the timepieces are bad, they don't have holes, little or big, where shit comes out. Michelangelo had fun cutting up bodies in the back room, but I'd rather deal with these little pieces of metal and clean water. And mathematics. Mathematics is really clean."

"It's also boring, unless you imagine getting answers when you measure, and do the math as just a metaphor for shit coming out of little holes. A rat eats some grain and pretty soon shit comes out of its little hole. You sit around here and gather data and put it all in some sort of mathematical digestive tract, and pretty soon shit comes out the other end."

"It's not shit. It's an answer."

"If you tell Father Pietro what you did, he'll say it's shit. Well, fecum. It sounds better in Latin. But listen, I'm with you. You gather information on your shit, I'll get smart on mine."

"You're Latin's lousy."

Salvatore ignored that. "So, how's this game coming?"

Galileo had been swinging the pendula of various lengths, timing with his clocks, entering data on the paper which Fra Basilio had found for him, with the quill and ink Fra Basilio had found for him. "Hah," he said. He had long lists of numbers, trailing down from various headings. "Game, huh? Shit, huh? Look at this."

Salvatore glanced at the lists. "You make very nice eights, and really good twos. It's very neat shit."

"No, no. Look at what it means!"

"Galileo, you have to tell me. What have you proven?"

"That how long it takes a pendulum to swing

from one end of the swing to the other and back again doesn't depend on how far the swing is. It only depends on how long the string is. A pendulum of a given length takes the same time to make very big swings as it does to make very little ones. I knew it. I knew it. My pulse told me so."

"I don't see how you could count on your pulse. You get so excited, it can't stay constant. It's an even worse clock than Pasquale's clocks."

"That's why I came here. And used better clocks than my pulse." Galileo thrust the data sheet into Salvatore's face. "Look. Just as I said."

"That's what you said you said? You have to explain all those numbers."

Galileo sighed. "Okay. Look. This row shows the period for a pendulum thirty inches long. And look. I pull it back just three inches and let it swing. Then four inches. Then five. And I time the period until it stops swinging. All the periods are the same, as long as the length is the same. If I make the pendulum shorter…here…the period is shorter, but always the same, no matter how far it swings. If I make the pendulum longer, the period is longer, but always the same, no matter how far it swings. I think this has something to do with Aristotle's Falling Down Law, but I have to think about it some more. What I also have noted is that the pendulum tries to get back to the height to which I displaced it. Never higher. As if it remembered how high it was.

Always just a little lower, but just a little., as if Aristotle is reminding it that it must stop, sooner or later."

"Which says that Aristotle is right, and you should worry about other things."

"Maybe. But someday I have to tell people about this. I can think already of other tests I could make."

"And what would that get you, besides trouble with Father Pietro?"

"I don't know. Maybe better clocks. We're always complaining about the clocks."

"No, you're always complaining about the clocks. Me, I'm complaining because it's hard to get out and visit the farm girls down in the village. They are not afraid of mortal sin, you know."

"We shouldn't even think about such things."

"We already are." Salvatore produced his own paper. "If you come with me, I'll let you use one of my poems. The village girls are especially unafraid of mortal sin if you read a nice poem and tell one of them that you wrote it just for her."

Galileo glanced at the paper. It read:

Oh pretty lady
Dark eyes so bright
Your charms like stars
In the blackest night

Oh lovely lady
Your cheeks aglow
The fairest prize
I'll ever know

He shook his head. "I rather show them my data page than this terrible poem."

"Michelangelo wrote poems, you know. To Vittoria Colonna."

"He probably went to Hell for that."

The two did go down to the village, after Galileo secured his data page in his secret space. Fra Basilio, in addition to building the secret room, had also found a way through the Academy wall for students who wished to do something besides study and pray. He understood the curiosity of young men, the questions about the souls and especially the curiosity about the bodies of the opposite sex. If they were to become trustworthy Catholic men they needed to get those questions answered. He knew that Salvatore would help Galileo along those lines.

Galileo and Salvatore were no longer boys, although they could hardly be called men either. They were essentially physically mature but had a ways to go culturally. This age varies from individual to individual, culture to culture and religion to religion. The culture of the Italian High Renaissance was considerably more confusing than the culture of the Dark Ages, at least when it came to women's bodies. In the Dark Ages, women's bodies were of no use until they were married. If they never got married, they were of no use at all. It was sinful to even contemplate them. Now, these bodies were being glorified. Botticelli had made the Birth of Venus look like the birth of Mankind itself. Of course, Botticelli could make a sea shell look like an engorged vagina.

While Galileo was a boy-man of the Renaissance more in the intellectual sense, Salvatore followed the new thought patterns more with his hormones. So, when Galileo sat under an olive tree to ponder the significance of his experiments with strings and weights, Salvatore wasted no time trying his poems on the nearest young females, several of whom were out in a grain field collecting seeds from whatever grass was growing there. Galileo observed. Finally, after Salvatore had convinced one of the peasant girls to join him in unpicked grass, where he thought he might offer her a different kind of seed, Galileo wandered over to the place where they lay and parted the tall cereal. He addressed Salvatore as if the girl weren't there. "Do you think this is what Fra Basilio had in mind?"

"Right now I'm not thinking about Fra Basilio. He's not here, and he needn't know what I'm thinking. Ever."

"That's true. But God knows."

"That's something I can address later. Are we not taught that God is ready to forgive our sins if we only ask?"

Galileo shrugged. "Then this is between you and God."

As if from nowhere, since she hadn't felt noticed, the girl suddenly asked, "What about me? Isn't it between me and God, too?"

"If you believe the stories in the Bible, it's different for you, just as it was different for Eve. If you believe the Bible, you and my friend Snake in the Grass here are reliving Eve's original sin. Women have a built in fault. But I suppose God forgives women as well as men for their sins. I suppose."

"I should have known not to get friendly with altar boys. Well, you and your friend can go to Hell for all I care." She humphed off.

"Along with Michaelangelo," Salvatore muttered. Galileo started to respond, but Salvatore stopped him. He made a chasing motion toward the girl's skirt, but stopped short of touching it. "We're not altar boys," Salvatore said, talking to the rear of the departing maiden. As he watched her disappear through the grass, he said, "We might as well be."

Then he turned to Galileo. "Do you expect me to forgive you for this? Do I interfere when you're swinging tops on strings?"

"Don't you see the difference? Before I swing those bobs I think hard. Is it what I really want to do? Is it forbidden? Will I sin against anyone? Against God? If I do, is it worth it? You knew what you were going to do was a sin. Against and with the girl and against God. Would it be worth it?"

"Probably. I will definitely think about it. After I get to do it. And if it feels as good as I think it will I'm pretty sure I will think it's worth it."

"Then I promise not to interfere again, if you'll forgive me this once."

Salvatore shrugged. "You're my friend. And I think you're going to need all the friends you can get. You're too damned honest."

Chapter Two

Salvatore, now fully endowed with a young man's hormones but constrained by his hidden ethnicity to be the most pious of students at Catholic academies, had learned early on that he was going to have to live a double life. This would have been quite difficult if he had had to go it alone, but he had Galileo as an example. In Salvatore's case it meant showing devotion to a religion whose only spiritual content for him was the music in the loft and the art on the walls, in the windows, and in the statues. Except the statues of a man hanging from a cross, which seemed merely morbid. For Galileo, his exemplar, it meant nodding to the church teachings of Aristotle as the prophet of natural phenomena, while all the while conceiving demonstrations that Aristotle had it wrong and , God forbid, Copernicus might have it right. Because Galileo was honest in his devotion to truth through physical proof yet also honest in his devotion to Catholicism, Salvatore committed himself to simple deceptions and, more importantly, to the friendship of Galileo. By the time he and Galileo enrolled at the University of Pisa as medical students, he was an expert at piety and love of the opposite sex. That's the reason he saw Bufana Giacometti first. But not last.

Excerpts from the memoirs of Salvatore Isaac Fishbeingnola.

I knew from the start that Galileo wasn't meant for medical school. One day we were to study blood and its properties, and as the professor droned on, repeating things that had been observed years before and never questioned, never proven, I could see in my friend's eyes the boredom, the skepticism, the disbelief I had heard him exclaim many times before.

This made me remember the night in the Boar and Goblet with our little group arguing with Bellarmino's little group. This was Maffeo Bellarmino, the younger brother of the current darling of the church, Roberto. The two Bellarminos were alike and yet different. Maffeo had light blue eyes – fish eyes. He was mean and egotistical by nature. Like many young men in Europe, he was a devout Catholic with no soul. His brother Roberto had dark eyes, the kind that hid the soul, the kind that bespeak passion in a temptress but reveal nothing in a priest. I had decided that Maffeo could always be trusted to do the meanest thing available, but Roberto was unpredictable. His piety was unquestionable. His use of that piety could be open to debate.

The subject of transubstantiation came up. Maffeo, quoting from his brother's writings, was certain that at the moment of consecration the wine actually changed to the blood of Christ. "Nonsense!" my friend exclaimed. "Wine is wine. It's atoms are those of wine before the consecration, and they are those of wine after the consecration. Any proof that wine is wine would hold after the consecration. Any proof that the substance is not blood would hold after the consecration. But nobody ever does these things. And the bread does not become the body of Christ. I could prove that too. But no one will let me. The bread and the wine are metaphors, one of the great poetries of our Mother Church. But they are bread and wine, when they begin and when they end." Bellarmino mumbled something about heresy to his cronies, but some of our friends, although wide-eyed, nodded.

I had only seen the elder Bellarmino once, when he came to visit his brother, but I knew from that meeting he would be one of Galileo's biggest problems. He would allow his holiness to shade his views and he would fall out with my friend. I didn't know then how powerful he would become and it didn't matter all that much to me, since I was determined to stay clear of any antagonism with anyone clerical. I had to remain completely apostolic, wrapped in the gown of the Mother Church. I always laughed (to myself) about this. It is very easy to appear holy when you're not, since big lies are so much easier than little ones. Poor Galileo. Remember, before medical school he had studied for the priesthood. He was truly pious, but he was so honest and sincere and intelligent that he often appeared, rather than apostolic, ridden with apostasy. And Roberto Bellarmino was a Jesuit. Galileo respected him for that. This made my friend give some benefit of doubt to Roberto's brother and Galileo's main debater, Maffeo. I wanted to tell him to stay clear of both of them but he would have

considered that small of me and not worthy of
argument.

***

On the other hand, my own problem was that
of confining my dalliances to lower class girls who
wouldn't judge my soul by my lack of a foreskin.
One such girl almost cost me my status, since, as it
turned out, she was an object of attention for Maffeo
Bellarmino. Maffeo the Unscrupulous had more than
his share of human shortcomings, but his shortest
coming of all was in the honor department. He must
have been vacationing in the Alps when the honor
concepts were being handed out. When he squinted
through those fish-eyes one could see dishonorable
schemes hatching. In a way I could trust those eyes
to be an almost infallible alarm system. I could never
establish such a system for Roberto. I think Roberto
felt that if one could say the Confiteor backwards in
Italian, Greek, and Latin, and had translated it into
Hebrew, where it's backwards anyway, he really
needn't have any honor, at least towards men.
Maffeo, on the other hand, had no idea what the
Confiteor was about, which made us even. Although
I had memorized the damn thing, I never really knew
what I was saying, nor did I care. It did help to make
one look pious, though.

But I was musing about Maffeo, and the girl
who almost brought us into opposition. Her name
was Bufana Giacometti. She had very large breasts. I
believe, had she been literate, she could have read the

Gospels with the nipples of her breasts. They must have been light sensitive, since they were always erect and seemed to be reaching for the sun. These things distracted from her other attributes, which is a shame. She had beautiful teeth, even and white, and those dark eyes I mentioned women should have. Any man would be taken with her. I was. Maffeo Bellarmino was. Galileo was, or at least seemed to be.

That is, Galieo noticed her as soon as I pointed her out to him. I didn't do that out of altruism. It was obvious that I couldn't be intimate with a woman who might be intimate with a Bellarmino, or for that matter with any woman who might be intimate with anyone connected with the church or with the university. I tried to confine myself to some Hebrew harlots who lived south of the city, and to the playful girls who worked the markets near the gates. They usually didn't have nice even teeth like Bufana, and none of them had nipples that could read, but a lot of them had very dark eyes. For variety I often went as far as Lucca, when I had the time. There were lots of acceptable girls there who weren't Hebrew harlots or plum peddlers.

I knew Galileo would be attracted to Bufana once he noticed her. It should be said that he had the young man's fondness for young women, but that was often buried beneath his love of mathematics and his inquiring mind. The fondness had to be dragged out of him, and forcibly pointed in the right direction. The force in this case was the mention of competition with Maffeo Bellarmino. There must be a God, for there is surely a Devil. It is the Devil who makes me do these things. I would later regret this trick, but at the time it seemed like fun.

It was fun because I could see the details building in Galileo's mind as he watched Bufana fill empty cups and flash those wonderful eyes at the customers. She had been hired by the barkeep that very day, and she was already the most popular serving girl. Maffeo Bellarmino was the first to pinch her buttocks. Several other students, and even some of the clergy there present, also partook of the buttocks fondling, but most lost interest when the fish- eyes of Maffeo turned stony. Only Galileo refrained from the early groping, and only Galileo was undisturbed by Maffeo's gaze. My friend never undertook a project without careful planning, without a great deal of forethought, without an expected outcome based on a sound hypothesis. He watched how she held the cups and which men received the most favorable attention. He listened to her accent, observed her walk, her gestures. While I was still watching the nipples of her breasts to see if they had eyes, he observed everything else. While Maffeo was fondling first her left buttock, then her right, Galileo was visualizing a test of what this woman would like. Once he had the details complete, he could proceed with the test. Strangely enough, the test included neither buttock fondling nor breast staring.

"Your hands," said Galileo in her accent, "are really quite interesting. Strong and supple. They are handling wine cups, but they seem more suited to mandolin strings." This from a man who thinks I write bad poetry to gain romantic favors. Of course I was surprised when Bufana actually blushed, if only slightly, smiled sweetly, and used one of her recently complimented hands to touch one of Galileo's own. This made Maffeo Bellarmino frown and make a grunting noise toward my friend. The noise made Galileo look around, first up, then under the table, as if looking for a stray dog or perhaps one of the boars after which the place was named. This mocking gesture made Maffeo grunt yet again, more angrily, which made Galileo look all the more assiduously. I don't know how long this cycle would have lasted, spiraled as it were, if I hadn't offered to buy Maffeo's next cup of wine. My friend gave me a reproving look, but I had successfully distracted Maffeo. Philosophic and theological arguments had ended for the evening. Soon, Galileo left with Bufana. Maffeo left with a grunt.

I would like to be able to describe what happened between Galileo and Bufana Giacometti, but I wasn't there, not for any of their liaisons, which were a few. I do know that Maffeo Bellarmino couldn't forgive Galileo for his humiliation, a fact which would come back later to haunt my friend. Still, when I think about those dark eyes and those even teeth and those all-seeing breasts, I imagine that relations between Galileo and Bufana were probably worth any small part they played in later troubles.

Chapter Three

Galileo's family had first moved to Florence where he could study for the priesthood. His head was too conflicted for such an endeavor. He was a Catholic, true enough, but the physical world seemed to contradict the spiritual world, at least if one tried to take both literally. Galileo could see only one literal reality, and preferred to let the spiritual world remain in the spirit or the ghost. A metaphor. It was the image behind the reality, and that could be painted or played on the church organ or left to Dante. The real world, on the other hand, could only be understood by real world activities, activities that might lead to conclusions that were non-conformist, even against the catechism. He and his family thought medical school would fulfill his intellectual as well as his spiritual ambitions. They were wrong. Galileo's spirit was drawn to the abstractions of mathematics, and mathematics became the means to explain the physical world. Galileo was inventing the scientific method, and he didn't foresee the trouble he would cause. He wasn't innocent, but he wasn't intentionally malevolent.

Excerpts from the memoirs of Salvatore Isaac Fishbeingnola.

I was never very good at math. Oh, I learned what I needed, mostly to do the little chemistry involved in my medical studies, and to do the rudimentary measurements we took on the human body. I could at least count to thirty two, which was the number of teeth some people had, at least when they were fully grown yet still young. I found it easy to count teeth and I must admit I never understood those early monks debating the number of teeth a horse had, as opposed to simply counting them. Through this example I could understand Galileo's need to actually measure things, to count what could be counted, to time what could be timed, to see what could be seen. I agreed with him on this. The one thing I didn't count was how many times I told him to shut up about it, to do his thing and to give the results only to like-minded people. At first, he took my advice.

He was leaving medical school, which was fine for him. His family wanted to be sure he had a profession, and there were only four things considered professional; the clergy, medicine, academics, and the military. He had already gone through the first two, and with that mouth of his he could never enter the military, so he had to seek academic employment. I knew it would be in mathematics. He was already doing tests on falling bodies, or, because he had lousy clocks, on my father's marbles rolling downhill. He needed to slow the bodies down a little you see. He was mumbling something about the square of time, and I humored him by nodding, as if I understood what he was trying to get across.

I hadn't the latitude given Galileo. I was going to complete my degree at Pisa and enter the practice of medicine. This would give greater cover to my father, who had become a great patron of the arts while passing as a Christian. I would not only be a doctor, I would be a good Catholic doctor. I would prove that by indulging in the practice of beating impotent men on the bare buttocks to set them on the road to virility. It was common knowledge that anyone not a priest, monk, brother, or friar, or anything to do with the church, had the duty to produce offspring. Impotent men could not do so, and that might be the work of the devil. Personally, I always suspected wine, or an ugly wife, or, God forbid, a lack of interest in women. While still a student I would beat no buttocks, and I personally suffered no impotence. In fact, my desire for women was so strong that I missed many lectures and demonstrations because of urgent trips to Lucca. I have no regrets. I got my degree, and the barmaids and whores in Lucca were worth the extra effort. I used an alias there, and found it amusing that the girls were always glad to see Salvatore Perfunctorio, the Catholic with the circumcised penis. One, who was something of an artist, even made a sketch of my organ. I have often wondered where that piece of art lies.

Certain things troubled me, however, the greatest of which was the turn my friend's life was taking. Oh, I couldn't contradict the interest in mathematics, what with his associations with Ricci and Del Monte. I was happy for him when he got the chair of mathematics at Pisa. I was starting the practice of medicine, my father was, although aging, still a vigorous force in the arts in town, and I had my secret liaisons well coordinated. Everything should have worked to put my mind at peace. But two things about my friend bothered me. One was purely mechanical. The other was much more ominous.

Galileo had shown that there was a way to delineate periods of time with much more precision than those relying on controlling water flow or falling weights. Yet almost all of his tests, or experiments as he was now calling them, relied on very good time keeping. The timing of the pendulum was a good example, and it led to a true paradox, which would have been amusing if it hadn't been so frustrating. In order to more convincingly prove that a pendulum had a fixed and predictable period, he needed a clock that kept time as well as a pendulum. Of course, such a pendulous clock had not yet been invented. Even my friend, for all his tinkering, had not yet got around to that particular mechanism. It's funny that later, when the telescope was invented, he would improve upon it dramatically, and put it to both practical and philosophical work. But, I digress.

The second and most important thing was his unhealthy interest in the discredited Monsignor Copernicus. This doctor and priest, this devout man of Krakow, this professor, had had the audacity to dispute the Aristotelian precepts which formed the basis for the explanation of the physical world, the world this side of death, as we perceived it. More explicitly, he had rearranged the Universe of Ptolemy, a universe defined with such beautiful mathematics as to guarantee the hand of the Devine in its concept and construction, Copernicus' rearrangement made the sun, not the earth, the center of the universe. This was bad in so many ways, undermining the Mother Church in so many ways, that it was patently unacceptable. Besides, it would require the earth to move and to rotate upon an axis, neither of which was evident nor palpable. That it was essentially heretical was proven by the fact that the good Copernicus didn't tell anybody about it until he was sure he was going to die; by that I mean sure that death was imminent. Copernicus had run out on his own hypothesis, leaving intellectual adventurers like Galileo holding the heliocentric bag. I felt the need for a long talk with my friend. I went to see him at his place, and began with a more banal subject.

"I see you're still fooling with Bufana. I don't blame you. She can stir a man's soul."

"What she stirs isn't my soul. But, to be honest, I don't have that great an interest in her. The fun with her isn't she. It's Maffeo Bellarmino. He's not bright enough for a decent debate on serious subjects. The only satisfying way to engage him is to steal a woman from him."

"That kind of trick should be beneath you. Toying with a woman just to make another man jealous."

"It is beneath me. I find that I often do things that are beneath me…I do Bufana when she's beneath me. Don't you do things beneath you?"

"Of course. But you know my circumstances. The women I put beneath me are beneath me, but that's because I can't afford to expose myself, as it were, in my own social class."

"Yes, I understand that. And you should understand my shortcomings. It was your god that subsumed all others unto himself. There is, therefore, a bit of the devil in even the most godly of people." Galileo laughed at his own wit.

"Don't burden me with 'my god.' I'm not sure about any of them. I humor the priests of Jesus to prevent a painful turn of events. I humor the god my mother says I should keep in my heart to humor my mother.. And believe me, a Jewish mother, if not humored, can be just as painful as an Inquisitor. And certainly more annoying."

"Poor Salvatore. But you'll be a good doctor, I'm sure. A good, Catholic doctor. It will amuse me to think of you whipping the arses of impotent laymen to help them consummate their marriage. To help them breed more little Catholics. And that begs another question. How many little Catholics do you supose you have fathered? With all your gallivanting you must have successfully planted a few seeds."

"I may have had some impact on the

population of Lucca. I try not to think about it. Most of those harlots wouldn't know who the father was anyway. Why should I?"

Galileo went back to his tinkering, and I watched for a while. Suddenly, he sat erect, his eyes focused on the distant wall. "It just occurred to me," he said, "that a strange thing happened. Apropos Maffeo Bellarmino. Now you have me wondering."

"What strange thing? Wondering about what?"

"It was just last night. At the Boar and Goblet, as usual. And, as usual, I managed to out charm Maffeo for the affections of Bufana. Now, out charming Maffeo is not difficult, and it has become easier as he becomes more frustrated. But last night was too much for him. He followed us from the tavern, accosted me, threatened me with a dagger, declaimed all sorts of outrageous things. I think he would have attacked me physically, in fact was about to, when two strangers appeared from the shadows. They were burly men, obviously ruffians, and I thought for a moment all three of us, Bufana, Maffeo, and I were in danger. But one of the men went straight to Maffeo and said, 'We don't like to witness violence, friend, so leave this couple alone and run about your business. Or else we will give you all the violence you can handle.' Maffeo was startled for just a moment, but realized they were serious and he was overpowered. He gave me a dirty look and went back into the Boar and Goblet. The men said nothing to me. They just went back into the shadows. What do you make of that?"

"A truly remarkable event." I thought that a sufficient comment and wished to say no more. I let my countenance remain impassive.

"That's all you can say. I'm saved from violence by the intervention of complete strangers, ruffians at that, and you say 'truly remarkable' as if you were tasting some vintner's latest wine. And you didn't think it was all that good."

"I'm sorry. You're safe, and that's what matters. I have no idea of the ambience of the scene, since I wasn't there."

My friend thought for a moment. "You see. That's what I'm trying to make everybody understand. Unless one actually observes something happen, something he predicted would happen if certain conditions were met, he doesn't know *why* it happened. He can only have a flimsy opinion about it. That's why I do these experiments. And I will do more of them in the future." He paused, and looked at me sideways. "Still, I have the feeling that you know more about this matter than you're telling me. Why else would you be so tranquil?"

It was too late for me to feign alarm. I must now admit that I had foreseen this kind of trouble with Maffeo Bellarmino. Lacking the talent to win an intellectual contest, and the charm to win a woman's attention, he would resort to physical violence. For Galileo's sake I had hired the two 'ruffians' as secret bodyguards against just this thing. I paid a high price for their vigilance and for their silence. Rather, my father paid a high price for their vigilance and silence. He thought the money I sought was to gain access to cadavers to continue my medical studies. It was to protect Galileo. And headed in the direction he was determined to take, it was going to get ever more costly. "As I told you," I said, "I wasn't there. As long as you survived without harm, I'm content."

I think he suspected me of having a part in his rescue, but he let the matter drop. He had succeeded in putting me on the defensive, so I couldn't deliver my lecture on how I thought he should order his life. Besides, I had my own ordering to do.

Chapter Four

Less than a year after he was appointed to the Chair in Mathematics at Pisa, Galileo produced the paper *De Motu,* in which he described his experiments with falling bodies, in which he advanced work done by Simon Stevinus, and disputed the theories of motion described by Aristotle and espoused by the Catholic Church. He had the courtesy to publish in Latin, which meant it would be read only by scholars and academics, and therefore would not rouse lay people to doubt the prescriptions of the ecclesiastical crowd. The chief point of *De Motu* was that motion is motion, not to be divided by some Aristotelian pie slicer into 'natural' and 'unnatural. This start of Galileo's great contribution to science was not helped by the publishing of yet another work by Giordano Bruno, an ardent supporter of Copernicus, albeit a metaphysical supporter. And Tycho Brahe, by watching a comet fly though the solar system, had already ruined the crystal spheres of the Krakow priest, although Tycho's apprentice, Johannes Kepler, was still trying to fit them into the regular solids of Plato. What a world. Astronomers were also astrologers; spiritual leaders blurred the actual cosmos with the supernatural. The earth couldn't move, the earth did move. The earth was the center of the universe. No, the golden sun, the life giver, the perfect orb, was the center of the universe. Galileo, little by little, was pushing the earth out where it belonged, and giving it a little spin in the process.

Excerpts from the memoirs of Salvatore Isaac Fishbeingnola.

My friend had studied with certain big names in mathematics, had shown his proclivity and inventiveness, and had astonished them with his ability to turn mathematics into machines that could be used to test hypotheses. He got his chair at the University at Pisa, and I began my practice of medicine. He tended to take up ideas which were, to say the least, controversial. I would have tried to forestall this tendency if I didn't have a few troubles of my own.

I had begun my medical practice, and my parents thought it time for me to take a wife. Because we were passing as Christians, this question of my marriage forced us into an agonizing conundrum. For my children to be Jewish, which is what my parents desired, my wife would have to be a Jew. If, however, I married a Jew, it would certainly come out that I was a Jew also. Since I had made all the sacraments of the Catholic Church so far afforded me, coming out as a Jew could put me in a very unfavorable position. All my acts to this point could be considered sacrilege, and for this profanation they had another ceremony: it involved a big wooden stake, a bunch of tinder, a person tied to the stake, and a torch. The next sacrament on the living list was marriage. If I married a Jew, and presuming I was never going to receive the sacrament of Holy Orders, the last sacrament, Extreme Unction, would be given me by the torch. I didn't want that. My parents didn't either, but they were more conflicted than I. They weighed having their grandchildren not be Jewish against me getting burned at the stake, and couldn't make up their minds as to which was worse. It was worse for my mother, since my father had always been at least non-observant and was a pretty good pretend Catholic. My mother was more tied to the Jewish heritage, considered me a good Jew, and wanted her grandchildren to be good Jews. On the other hand, she didn't want her good Jew son to be burned at the stake. Personally, I had no trouble agreeing with the last consideration. I had been a non-Jew and a faux Catholic for so long that religion was of little importance to me. I was fine with marrying a nice Italian girl with dark eyes like Bufana Giacometti. She wouldn't even have to have eyes in her nipples.

While I was dodging potential clerical enemies Galileo had leaped ahead of me. He already had a bunch. It wasn't so much his work and his writings, yet. He was friendly enough with Pope Urban the Seventh. It was the Bellarminos. And Roberto was in a very high place. I had warned Galileo about this, but I was actually more hindrance than help. Here's why.

I was visiting my friend in his place of experiments, and we got to talking about the nature of motion, which didn't interest me very much, but interested him immensely.

"Why can't I make them understand?" he asked rhetorically. The 'them' referred to almost everybody, but the 'them' that counted were some Jesuit scholars and about half the natural philosophy community.

"Perhaps if you make *me* understand I can help with that question."

"And what is it that you don't understand?" He was mocking me. He knew I had read *De Motu*, and the implication was that if I read it I must be smart enough to understand it. He waited.

I wanted to be honest. I always wanted to be honest with my friends. Unless the truth would hurt them needlessly. That wasn't the case here. "I see nothing wrong with Aristotle. Things go up and then fall down. It is natural for things, in the earthly realm, to fall down. And stop. That is natural motion. Things fall down and stop. If they move sideways and don't stop, it is unnatural, because we must push them to keep them from stopping. We just know this."

"And what else do we 'just know'?"

"That big things fall faster than little things."

"And how can you tell that?"

"You just know by looking at them."

"What about dropping them?"

"You could do that, too."

"And the big things would fall faster?"

"I guess so."

"Well, I guess not. And Stevinus did some experiments, and I did some better ones. I did this off scaffolds here in my studio, and I did it off towers. If the objects are of the same material, and the same general shape, then size or weight make no difference. I have done this many times. But here, let me give you a quick demonstration." He produced from a table top a large, round melon. In his other had he picked up a smallish grape. He climbed on a chair and pulled a stand close. The stand had a basket he could just reach, in which he placed the melon and the grape. "You can see that when I pull the slide open, the melon and the grape will start their fall at exactly the same moment. Yes?"

"Yes."

"I will let you judge which reaches the floor first."

I nodded that I was ready. He triggered the mechanism. Out came the two fruits. Without question they hit the floor simultaneously. I knew he wanted my honest response. "They fell with the same speed," I said.

"They fell with the same *change* of speed," he corrected me.

"Whatever. It is still counter-intuitive."

"Not if you think about it. Picture this, if you can. I drop a very heavy object, and at the same time drop an equally heavy object, tied by a string, to a very light object. Are you with me?"

"Yes."

"Now, if the heavy object and the light object are tied together, they must weigh more than the heavy object alone. Yes?"

"Yes."

"So, if Aristotle were correct, they should fall faster than the heavy object alone. But if the light object is tied to the heavy object, it should hold the heavy object back a little, so the two of them, although heavier than the heavy object, should fall more slowly than the heavy object. Is a such paradox intuitive?"

I smiled a willing smile. "Of course not. But your experiment is even more impressive. One thing troubles me, though. You said it was not the same speed, but the same change in speed that they experience. What does that mean?"

"It means they gain speed identically. I have done experiments with rolling balls, to slow down this gain, so that I could measure its nature more accurately. This, in turn, has led to some very important ideas, certain abstract principles behind the workings of the real world. Let me ask you this. If you tried to slide a smooth, flat rock across a mound of sand, how far would it go?"

"Not very far."

"If you slid it across the marble floor in the Doge's Palace, how far would it go?"

"Much further."

"And if you slid it across a frozen lake, with the same push as the first two cases, how far would it go?"

"I don't know. All the way across, I suppose."

"So, this same "unnatural" motion, propelled by the same force, seems to have different stopping distances for different surfaces. But, if it is natural for it to stop, why does it have such a hard time making up its mind exactly when to stop?"

"I don't know."

"Yes, you do. It is because something is

making it stop. Just as it took a push to make it go, it takes a push to make it stop. And the sand pushes harder than the marble, and the marble pushes harder than the ice. And if nothing pushed it, it would never stop, and a person in another land would see it pass by and say, 'That stone has natural motion, because nothing is pushing it and it keeps on going."

"But there is always something to make it stop. To push on it, if you will."

"Exactly. That's why it stops. And that's why the moon doesn't. Because there is nothing pushing on it to make it stop. God gave it a push to get it started, and nobody has interfered since."

"But why does it circle the earth?"

"I don't know. Someone, I wish it were me, will figure that out. But this leads to very important things, all learned with my rolling balls. It seems that when one rolls down an incline and then up another, it will always reach about the same height from which it started, no matter how steep or shallow the incline. This says to me that if I don't make an uphill after the downhill the balls will always be trying to reach the height from which they began, and will never stop rolling. So, yes, down might be a 'natural' motion, but up isn't. And, since sideways is a result of down, it's natural too. What I'm really saying is that if an object is in motion, it will stay in motion until something, something physical, stops it. I call it inertia. This is much different than Aristotle's impetus. Impetus has some fixed quantity and runs out. Inertia requires a push to change it." He shook his head and smiled. "This is somehow related to the fact that all objects, big or small, fall with the same changing speed, but it's not yet clear to me how. It has important implications, though."

"Which are?"

"Well, anti-Copernicans say the Earth can't be spinning because we could feel it, and, besides, we would know it by dropping something from a high place. If the Earth is spinning very rapidly to the East, then if one dropped a ball off the West side of the bell tower, the tower would leave the ball behind as it fell, and the ball would fall well west of the base. But the ball doesn't. It falls straight down, as Aristotle would predict. Why is that?"

"You tell me."

"Because the ball is moving to the East with the same speed as the tower, and even though you let the ball go, there is nothing pushing against it to make that speed go away. So, all the way down, it will keep that same speed to the East as the tower, and hit at the base. Not because of what Aristotle thinks, but because of what I call inertia. Not because Aristotle was right. Because I am."

"So you think you're smarter than Aristotle."

"No. He was a very smart man. He just

didn't play with toys enough."

I was by now mentally exhausted. These ideas which Galileo had might be right, but they would take a great deal of consideration. And observation. And putting the two together. If he were right, it meant more than just believing Copernicus. It would change the way we would look at the universe. Even the way the Church looked at God. I had only a vague inkling of the import, but such deep subjects were not why I had come to visit. It was a good time to change the subject. "So, outside of rolling balls and dropping cantaloupes, how is your social life going?"

I received, for that question, one of those sideways looks. "I know you ask as a friend," he said. "But have you heard anything about me."

"Why, no." This was one of those times when lying to a friend was for his own good.

"It's just strange that you bring that up just now. After last night."

"Last night? What happened last night?" I knew, of course. My hired ruffians had reported to me. They were indeed seeking a bonus for an extra good job.

"Well, I left the Boar and Cup with Bufana, Maffeo Bellarmino displayed the usual dismay. He didn't know we weren't headed for an assignation.

No. I wanted to tell Bufana that we could no longer have our liaisons. I had other interests. I didn't say what they were. When I did tell her, she was crushed."

"So you ended your affair with Bufana, of the dark eyes and...." I left the part about eyes in her nipples out. It would start an argument comparable to the one about transubstantiation.

"It wasn't an affair. I'm ashamed to say it was more of a prank played on Maffeo Bellarmino." He shook his head and his eyes saddened. "Poor Bufana. I wasn't fair to her. But she wouldn't let me finish my apologies. She rushed back into the Boar and Cup. I turned and started for my house, but before I got very far I heard footsteps, running footsteps behind me. I turned to find Maffeo charging down the street, and as he got closer, I could see rage on his face and murder in his eyes. I prepared to defend myself, but I must admit, I was fearful for my life." He paused, and gave me that quizzical look again, as if considering that I might have a part in all this. "And then, just as before, two ruffians appeared from the shadows, chasing after Maffeo, closing in on him. Before he reached me they caught him, threw him to the ground, and gave him a thorough thrashing. I was now afraid for him, rather than myself. I told them to stop, or they would kill him, but they didn't listen to me. They didn't stop until a crowd of men from the Boar and Cup finally arrived at the scene. The ruffians ran off, leaving me standing over Maffeo and pleading for help for him. Some of the men carried him back to the tavern and said they would fetch a doctor. No one said anything more to me. In fact, they left me alone. Alone and wondering." Again, that look. "The providence of two ruffians coming to my aid? Twice? That's what left me wondering. Do you know anything about this?"

"Me? Of course not. This is the first I've heard. My god, you should have come to my house immediately. We could have shared wine and relief. That's what friends are for. Why didn't you?"

"I needed some time alone. I wanted to put it out of my mind as quickly as possible. After all, I was alive, Bufana was out of my life, Maffeo was attended to. What more could I do about that? There is so much on my mind, like the mathematics of falling bodies. It's starting to jell. That's why I needed to rid myself of the distraction of a woman. For just a while. But I can't get it out of my mind, mainly because there is some unseen hand in this. The ruffians were more than coincidence. If you say you're not the guardian angel, I believe you. But somebody is."

Once again I was in need of changing the direction of the conversation. I went with a sure thing. "Well, you may never know. But tell me, are you working on anything besides rolling balls?"

Once again he was enthusiastic. "I'm trying to find a way to measure temperature. I have several concepts in mind, but none seem very accurate. I haven't started to make any instruments yet."

"I'm sure anything you come up with will be very good."

"If by good you mean better than a mother's wrist, maybe."

Chapter Five

In 1592, Galileo got a better offer. For more money and a little more freedom, he moved to Padua. He took with him the finished thermoscope, that better thermometer he had mentioned to Fishbeingnola. As he had told Salvatore, it was a little better than a mother's wrist, only because it had numbers on it.

Padua had another advantage besides high salary. Although Maffeo Bellarmino had ceased all forms of intercourse with him, Galileo still felt uneasy about the man. He had, as intended, stopped seeing Bufana. In fact, he had given up women for that period, intent on his writings and experiments. He just knew he could convince the world that the earth circled the sun, and that there was more to motion than intuition.

Excerpts from the memoirs of Salvatore Isaac Fishbeingnola

My parents finally conceded that, for my safety, I should marry an Italian girl, a Catholic. We could work out the Jewish mother thing somehow. After all, we were all secret Jews, so she could be one, too, the difference being that it was not only a secret from the world but a secret from her. She would be a really secret Jewish mother, which pleased me immensely, since it meant she wouldn't act like a Jewish mother. Our children would be faux Jews, which made us even, since I was a faux Catholic. Making a personal observation, I found there to be few true Catholics, so I didn't feel so bad about faking it. Maybe Augustine was a true Catholic, but very late, and grudgingly. Ironically, my friend Galileo was as close to a true Catholic as anybody, but people like Bellarmino (the pious one) would contest that.

This wasn't the gist of my problems, however. I use the plural, because my life had two main problems. First, there was choosing a wife. There were so many young women available in our social circle, who must marry in class, so the field was wide. Most, however, were either homely or slow witted or both. I found one, Gioia Gnocchini, whose father was the king of dried pasta in Pisa. Gioia was bright and witty, could play the mandolin, and while she didn't have the dark eyes and all seeing breasts of Bufana, she was quite pretty and deliciously chubby. I have been, since childhood, drawn to voluptuous females, and this is another reason for my avoidance of a real Jewish mother. Jewish mothers, purposefully or not, somehow taint voluptuousness. I don't know why that is. Gioia, on the other hand, juggled her voluptuousness right before your very eyes, with all the skill of a professional. In truth, I fell in love when she was pouring tea for me, locking me with her radiant smile and deep blue eyes, and nearly boxing my ears with her ample breasts. Ample, mind you, but not literate. Incidentally, her given name was the same as my father's nickname for my mother. She was instantly approved by my parents.

Gioia turned out to be one of the best and most important decisions of my life. With each day I knew her I loved her more. Although she had grace, education, and beauty, she had the humility to treat everyone as a fellow human being, with a gentility and kindness I had, to this point in my life, seen in no one else. She made me realize that one could learn some of the best of life's lessons from example, simply by paying attention. After Gioia, I often averted anger at a person or situation just by thinking of her. Not always. Sometimes anger is necessary.

Wedding arrangements were made, which brought me to my second problem. While I had all but given up on Judaism, letting the Mosaic laws slip into antiquity where they belong, I was not comfortable with the graces of the Mother church. It was one thing to take communion with a group, and mumble the required Latin phrases during mass. It was quite another to keep a straight face and a pure soul while performing the sacrament of marriage, alone (but for the bride and priest) and naked (emotionally) before the world. Would I cry out, "Just make the pronouncement, for heaven's sake. I love this woman and will never forsake her. I have better things to do with my energies than chase after women, now that I've found one good one. That's all anyone needs to know." I would sweat and tremble throughout the entire ceremony. I wasn't worried about going to Hell. If there were such a place, it was a foregone conclusion that I would go there. Dante hadn't invented a level for me. It was up to me to create that Hell for myself, heading there in the most enjoyable manner possible. I wanted to arrive there having last seen Gioa's face and voluptuous body, not a scowling monk with a torch.

As it turned out, I had another, much larger problem. Maffeo had discovered, somehow, who financed Galileo's protection. He knew it was I who was the sugar in the bodyguards' tea. My hired ruffians had mentioned their employment to ruffian friends, friends who had friends who often worked for Maffeo. So this information went full circle and eventually came back to me, but by the time I knew that I was the object of Maffeo's revenge for the beating he took, he had already tried to extract that revenge.

Galileo had accepted the offer to teach mathematics and further his studies in astronomy and motion at the University in Padua. He felt it was a more prestigious position, and perhaps a chance to leave the distractions in Pisa behind. For my part, I could see a move to Padua as a chance to further imbed myself in Christian society, to further shield myself from Inquisitive eyes. I would even change my name. Oh, not Fishbeingnola, which was a very common Italian name. Just my middle name. Before I could make the move to Padua, however, Gioia brought even more joy into my life.

A scant nine months after our marriage, my darling Gioia presented me with a son. He was chubby, like his mother, and healthy - full of life. We named him Vito. Vito Abraham. Vito Abraham Fishbeingnola. I was so proud and happy, and remained so for a few days.

It became apparent in that first week that Gioia was not recovering from the delivery in a normal fashion. Although I was the most modern of doctors, I couldn't make a certain diagnosis. If she had been a dog I would have known exactly what to look for, since we had vivisected pregnant dogs in medical school. A human being was more of a mystery, however. I found that I couldn't sleep, worrying about solving this problem and about the care of the newborn Vito. All the while Gioia faded and that made all problems but her salvation disappear from my mind. I had to see to Vito's care, but I oouldn't leave Gioia's side. I was a doctor, but I felt helpless. I couldn't think of medicine, except how much I didn't know. .

I did know I couldn't take the place of a newborn's mother. I had no idea how to care for an infant, so I put him in the care of my mother-in-law and a wet nurse while I worried over Gioia. My own mother felt slighted, but finally realized that was the right thing to do. I must make an admission here. There were brief interludes when I actually blamed Vito for his mother's plight and was glad he was not near me or my parents. It was during my submergence into one of these irrational thoughts that

Gioia's humanity came to my rescue. It was a simple epiphany - simply apply the philosophy of compassion I had learned from Gioia. I looked at that baby and vowed he would never be without unconditional love, for his sake, for his mother's sake, and for mine.

Is there such a thing as Kismet? Or Fate? One often wonders why things unfold the way they do, why there is often a hypothetical question, a what if, an action that would have made certain events have different consequences. What if I hadn't brought Vito to my mother-in-law's house? What if I didn't have to divide my time between my two greatest loves, my son and my wife? What if I had been napping next to Gioia, as was my wont when I was there looking after her? What if I hadn't gone to my mother-in-law's house to coo at Vito? What if?

I will never know if my visit there saved my life or guaranteed Gioia's death. For, while I was gone, my home was consumed in flames. The tapestries, the furniture, the wooden cornices and ceilings, the long wooden beams that supported the upper stories, all turned to ashes in a few moments. The servants on the lower floors managed to escape the flames, and one agile maid leaped safely from the second story. Everyone else in the house, including the mistress of the manor, was caught up in the holocaust. A neighbor came screaming to my mother-in-law's door.

I returned to the scene of devastation to spend

hours morosely staring at the still smoking ruins. It seems so strange now. Even though under normal circumstances we would have moved to Padua without a second thought, the smoking shell etched itself into my emotions, claiming a kind of sanctity that I would never have felt if I had simply sold it to a stranger. Ever after that, the smell of smoldering wood would bring me back to those moments of despair.

In short order I learned from the same ruffians who had given a beating to Maffeo and who had unwittingly given me up as their employer, that it was Maffeo who arranged the arson. I would like to say I swore immediate revenge. The truth is, I was in no state to swear anything.

For a few months following the disaster I was inconsolable. I couldn't eat, I couldn't sleep, I couldn't talk to even those closest to me. I reached for the wine and it didn't help. I shunned society because I couldn't stand the slightest hint of happiness. I cursed Galileo as if it were his fault that Maffeo had ruined my life. And I couldn't even look at my son, my proudest achievement, because his chubbiness and the shape of his eyes reminded my so much of his mother and of my loss. For just a while I didn't care if I lived.

It was not a priest who pulled me from my despair. It was not Jesus, and it was not the God of the Old Testament, who didn't do things like that anyway. No, it was my father and mother, Filipo and

Giubba, who tended me without intrusion into my grief – yielding only love and reassurance, reassurance that life would go on, and that Gioia, the effervescent Gioia, would want me to go on with it. They finally restored my resolve to move to Padua and start life anew. My parents, Gioia's parents, and my friends, all encouraged me. I would have a motherless son but not be that far from two doting grandmothers. He and I and a good nanny could take on the world. I would be a good father and a good doctor, and there was a bonus. As I emerged from my self-imposed dungeon I had realized the obvious – Galileo wasn't to blame for Maffeo's villainy. I was. No. Maffeo was. Now I felt the urge for revenge, but that couldn't be done with public justice, since I would never be able to tie the arson to Maffeo. I would wait. I had better things to do anyway: reestablish my close relationship with Galileo, who was already in Padua, establish a practice there, and be a good father. Revenge would come if there were ever an opportunity for an appropriate action. In the meantime, Padua would be my new life.

My ruffians did me the favor, for a small price, of learning the details of Maffeo's plot to assassinate me. I learned what my meddling in Galileo's affairs had cost me. He had not asked me to meddle. He would have disapproved if I had proposed providing protection to him. No, this horrible event came back on me, but it was consummate injustice just the same. For a few bruises to body and ego, Bellarmino had commissioned murder. Yes, I would probably seek revenge, but not soon, and not blatantly. I knew it would be in the back of my mind for as long as I lived.

***

I found a very nice abode in Padua, where Vito and I could live upstairs and I could establish my practice downstairs. When I hung my shingle, it reflected my very clever name change. It read:

Salvatore Omar Fishbeingnola - Doctor of Medicine

Chapter Six

Excerpt from the journal of Galileo Galilei, May 14, 1593

I sometimes feel frustrated in my quest for provable truth concerning the nature of motion and the nature of our Earth's place in the universe. It has become obvious to me that advancement in knowledge about the nature of things requires advancement in knowledge in all subjects. If one is to demonstrate proofs by physical experiments, one must be able to measure ever better, to see ever better. This requires better clocks, and the study of the nature of light itself. My amanuensis and I tried to measure the speed of light today, without the better clocks I yearn for. My conclusion: If the speed isn't infinite, it is far greater than I am capable of measuring. I can't help but feel that the lens makers will invent something that will aid in this study, and in the study of astronomy itself.

Excerpt from the Memoirs of Salvatore Omar Fishbeingnola, ca. 1599

My practice in Padua flourished from the start, and my clever name change had thrown Maffeo Bellarmino off my track. Omar, after all, wasn't a biblical name. Neither my family nor myself had been threatened by him since the Pisa fire. Maffeo seemed to have forgotten about Galileo too, although we didn't know that he had a continuing communication with his brother, and that his brother Roberto was the cause of the occasional notes of disapproval which my friend received from the Pope. Galileo felt he was engaged in pure intellectual endeavor, and was satisfied of this when he published *On the Mechanical Sciences.* How could he give regard to admonishments from the Pope when they really didn't apply to him?

As I have said, I didn't wish to ever marry again, but I found I wasn't bereft of my need for intimacy with the opposite gender. At first I took up my old habit of going to a neighboring city for romance. Venice was my city of choice, the clothes of a soldier my disguise. Many a Venetian barmaid was courted by Captain Gugliermo Gonzaga, commander of the guard for the Duke of Ferrara. I even had the letter GG embroidered on several tunics.

Somehow, Vito thrived without a mother and I thrived without a wife. I felt no need to marry again. I had been married nearly a year before I lost my beloved Gioia. That loss was almost more than I could bear. I didn't want to take that chance again. Still, the practice I aimed for couldn't thrive as long as I was a bachelor, a handsome enough bachelor of good family. You see, I aimed to specialize in women's problems, wealthy women's problems, and most of the married women who would be my clientele had husbands who were at the very least wary of me. I didn't want to be limited to pregnancies and maladies from the waist up. I needed female protection from jealous husbands. Of course I would have had allies if the women had lovers in addition to husbands, but I couldn't see how such allies would do me any good. "My good sir, I am not doing anything improper with your wife. Just ask her lover." No, I needed a female ally.

I chose to attend mostly to women of the higher social classes. Their problems were usually minor, except in their own eyes. They were problems that would correct themselves if left alone. However, if I could be sympathetic and discreet I would be considered the doctor of choice for these women. After I established my impeccability, the practice became rather easy work with great pecuniary rewards. As a bonus, I had gained financial independence without yielding most of my personal pleasures.

My savior came in the form of a Swedish woman. Anna Bolen. I would never have guessed that she could captivate me, since she was blonde and blue-eyed, well-shaped but not voluptuous. Just the antithesis of my Gioia. I had advertised for a nurse who might also act as Nanny, and this Swedish vision miraculously appeared. As soon as I saw her I saw my protection against jealous husbands. She would be in the examination and consultation room with me when I had to deal with delicate problems. I would even encourage husbands to accompany their wives to my office, at least for the first visit. Some of them insisted on coming every time, until I had Anna adopt a more severe appearance. With Anna in situ, no one would question my professionalism. The only negative, and a small one at that, was that I had to hire yet another woman to be nanny, housekeeper, and cook. The bigger positive was that Anna became, in addition to nurse, my lover, and I could throw away all those silly tunics with GG embroidered on them. My life, absent the trips to Venice, became even more manageable and pleasant.

Anna was a wonder. When she converted to Catholicism (willingly I might add, unlike me) she was forced to leave her very Lutheran homeland and travel far enough south to avoid the war that was looming between the forces of the Swedish king and those of the Pope. She packed what few possessions she had and just left. She never detailed the exact circumstances, but I gathered her departure was sudden and unannounced. She had, as many strong minded young women do, decided on her own spirituality, and saw it residing in the Catholic Church. Her Lutheran family disapproved and treated her poorly in an attempt to make her see the corruption of the Papists. They lost her as a result. Her approach to Catholicism reminded me of Galileo. She was a true believer yet saw the forgiving God who absolved sins through an intermediary in the confessional as the most rational kind of God, one that takes into account the imperfection of man and allows him to sin, as long as he is later sorry. This made her absolutely refuse to consider marriage. She wasn't about to give up her freedom and a good job with sexual bonuses, or ruin our delightful relationship. She would pay in Purgatory, but achieve Heaven. For this I loved her. That, and she was a very good nurse and lover.

***   .

So it was that I dealt almost exclusively with women in one way or another, and my friend was involved almost exclusively with things mathematical and their translation into things mechanical. We were a couple of dull fellows, so when I suggested a sojourn into the Tyrolean Alps for diversion and recreation, he agreed. It would be easy for both of us. Galileo had a girlfriend. Although they were having conjugal relations, they had not officially conjugated. He was free to travel.

I loved Vito and I loved Anna but I also loved hiking and wandering in the Tirol. I had been retiring to that lovely region for years, in fact ever since moving to Padua. After my first visit, I had begged, pleaded, and cajoled Galileo into accompanying me there, just once. I felt that if I could get him there he would be soothed and refreshed by the place, just as I was. It was a place where one could be alone, where one's soul could find itself, where, even if one ran across the occasional native he might not speak because the probability was that person spoke German, and German speaking Tirolians were notoriously close-lipped. We Paduans called this getaway village Commune de Tirolo, but the natives called it Gemeinde Tirol. Whatever the village was called, the little inn just outside the village, on the ankle of the mountains that surround the valley, embraced me as strongly as a woman could. The first time I realized this I was shocked, thinking I wasn't myself, that I had lost my manhood, even my sanity. But I came to expect this feeling, and hoped I could pass it on to my son Vito Abraham, and, to pass perhaps some small semblance of that feeling to Galileo, so that he would let his work vacate his mind momentarily. I knew he was a genius, a great man who owed mankind the proceeds of his talents. Still, he could relax just a little. It would do him good to be in a place where man's progress from cave to palace could only be detected by the Castle Tirol on the top of the hill to the North. So I begged him to go there with me so much he finally agreed.

***

It was two carriage rides from Padua, with a stopover in between. The roads were somewhat rough the whole way, but as we climbed into the Alps the scenery made us forgot the bumps and knocks. The smell of the alpine foliage drained worries better than several glasses of wine could ever do. The sights of the changing shades of green mixed with the sprinkles of yellow and red from flowers amplified this medicine, at least for me. Galileo? I think it worked for him also, although he seemed lost in thought as usual. We conversed a bit about nothing important. Once in the mountains, even on dangerous roads, the ride was dellghtful.

We were greeted at the inn by Geronimo, the blind innkeeper. Geronimo was bilingual - although his first language was German, he spoke decent Italian. He was happy to see me through my voice, one way blind people see. He got around the inn and its grounds as if he were sighted, and he knew his regulars sometimes even before they spoke. Perhaps by their walk on the wooden floor, perhaps by the smell of their pipe, perhaps by the sound of an accompanying dog. I never made reservations there, so after greeting Geronimo my first question was about the availability of room.

"I have two good ones for you to choose from." He stopped. "Or perhaps you and your friend would like them both. I don't believe I've met this gentleman."

I wasn't surprised that the blind man knew I had brought a stranger to him. "Forgive me, Geronimo. This is Galileo Galilei, a dear friend. He's come to go hiking with me." Galileo extended his hand and by the sound of his sleeve's rustle Geronimo found it.

"It is a pleasure to meet you, Galilei. My sister has read me some of your works." As if he could see the surprise on Galileo's face, he said, "Oh, yes. We get books here in the Tirol. Interesting books, like yours." He fumbled under the desk for something, then produced *On the Mechanical Sciences*. "See?" He smiled. "This is going to be a good season. I have several newcomers. You, of course, are the most illustrious."

We took both the rooms he had available, threw our baggage into them, and adjourned to the great hall of the inn. I was anxious to start a hike even then, but it was too late in the day to go where I wanted, and I didn't want to give my friend a wrong first impression. I would let the little town of Commune de Tirol speak a welcome. We would just wander down to the village. Galileo still looked a little distracted. "I want to make this as relaxing as possible," I said to him. "If you'd like to sit for a while and enjoy the ambience we can take a short walk this afternoon around the village. Or, if you're up for something longer there are several nice tracks that start nearby, with various lengths and durations."

Galileo shrugged. "No, I don't want to sit here and vegetate. The ambience, as you call it, is out there. Up there." He nodded toward the mountain. "But I'll settle for a walk around the village for today." I jumped up, but he wasn't through talking. "Tell me," he asked Geronimo, "if the inn is full, are all the people already out wandering the hills?"

"Yes, probably. That is the main reason people come here."

"Interesting. Would you know some of the regulars?"

"I think I would."

"The fellow who came into town on horseback about the same time we got here. Is he familiar?"

"I don't know whom you are talking about. I haven't been down to the village in days. My sister is there now. If you run into her you can ask her about this contemporary arrival. She would know. Especially if he came into town on horseback. He's probably a local. Can you describe him?

"Nothing special about him. Didn't look Tyrolean. Didn't have Tyrolean clothes, although you don't either. It just that I can't help thinking I've seen him someplace before."

"I can honestly say I'm sure I haven't," said Geronimo with the shade of a smile. "Did he have baggage?"

"Nothing like our bags."

"He doesn't sound like our normal visitor. Most come for a stay and have baggage. My sister is your best source of information on this. It's a small village. I'm sure you'll run into her."

Galileo dropped the thought with a little toss of the head. "So," I said, "if you're ready, let's have a walk and see if we can find Geronimo's sister. Or, failing that, someplace with a good bottle of wine."

He consented by starting toward the door.

***

The village, perhaps a mile downhill from Geronimo's inn, was typically small, with just enough amenities to make it a seat of local social life and whatever governance that might be required to make life manageable. It had a small chapel which housed the usual statue of Jesus on the Cross and a rather poor painting of the ever weeping Mary. The painting was poorly positioned and lacked perspective, so that Mary seemed to be weeping over spilled milk rather than the dying Jesus. The chapel had a few benches and plain windows. I had often thought to ask my father if he could perhaps put together some stained glass for the place – maybe a better Mary, one actually looking at Jesus. I never got around to asking, since I was fonder of Geronimo's house than the chapel.

There was also a tavern, a bakery (which also made pasta), a stable, at least one other inn, a three table restaurant, the expected town fountain, and a few little businesses, like seamstress and teamster and other functionaries. One had to know who they were. The businesses weren't signed and the people in each did several things besides hauling firewood or sewing clothes. It should be said that there were actually two or three cobbled streets, which made for two or three more than most Italian villages of that size.

Galileo and I chose to stop at the restaurant,

since one of its three tables was outside. We ordered
a bottle of the vino de tavola, since that's all they had
anyway.

We had only taken our first few sips when I
spotted Gina, Geronimo's sister, coming along the
cobbles. She was coming from the direction of the
vegetable woman, swinging sacks in both hands. She
was a big, healthy girl, strong and adept at fetching
groceries, cooking, and cleaning at the inn. She was
not unattractive, since she kept herself well. I will
admit that we had had a few romantic entanglements
before my marriage. We ended those amicably. I
still remember the smell of her sweat, for me a very
pleasant thing. I waved to her.

I had always liked the timber of Gina's voice,
and it seemed even more throaty when she said,
"Hello, Sal." Even Galileo straightened a little in his
chair. When this peasant maid spoke, she was that
engaging. "Here for your soul, as you say?" she
asked me. She was already appraising my friend.

"Quite so, Gina. I'd like you to meet the man
I've spoken so much of. This is Galileo Galilei."

"Ah," she said. "The famous Galileo." They
touched hands.

"Famous?" Galileo said. "Salvatore talks too
much. I am just a professor and observer of things."

"But the things you observe. And what you

say about them. It makes me glad I can read. I feel
connected to the things only gods could ride before."

Galileo looked at me, then back to Gina.
"What is a woman with your mind doing shopping for
vegetables in Commune di Tirol? Wait. I forget.
Society tells us there is no such thing as a woman
with a great mind."

Gina laughed. "First, I love it here, for the
same reasons Salvatore comes here as often as he can.
And I had to learn to read because of my brother. My
brother's blindness was my blessing."

Galileo nodded. "Unintended consequences.
I understand that quite well. Most of my discoveries
have unintended consequences. Or at the very least,
unforeseen consequences. And that's one of the
reasons I'm here at Salvatore's urging. To try to
forget some of the consequences at least temporarily."

"Whatever the reason," she said, "it's nice to
finally meet you."

Gina shared our wine. The rough oak table
had been stained many times by the house red wine,
and we spilled a few drops ourselves. Gina drank
wine like a man, with hearty draughts and a burp or
two. The conversation remained light until Galileo
asked her the question he had asked her brother,
concerning the man who had arrived
contemporaneously with us, but on horseback.

"Yes," she said. "I saw him. He's staying with the Berglosi's. I didn't talk to him. He left his horse at the stable and went straight to the room they rent."

"Is he a regular in Commune di Tirol, or a local"? I asked.

"Neither. "I've never seen him before." She paused. "Something about him. Rough clothes? Heavy build? Hard face? I don't know. Not our usual city folk come for peace and quiet. I shouldn't judge, since I really know nothing about him."

Galileo looked at me. "Unintended consequences?"

I shrugged. We walked back to the inn with Gina.

Chapter Seven

*Continuation of excerpts from the memoirs of
Salvatore Fishbeingnola*

After breakfast the next morning I had Gina
put some victuals and wine in a rucksack. She was
such a sweet woman and she never seemed to tire
from the mundane things an innkeeper had to do. She
had been up early, milked a cow, baked bread from
dough left to rise overnight, gathered a few eggs,
cooked breakfast, and then cleaned up after herself.
All these before we even got started, and she was still
more cheery than Galileo, who seemed to be brooding
over the unknown horseman. At least, that's what I
thought caused his morose behavior. I finally said,
"Look, my friend. We are in one of the most
beautiful places in the world. It is ours to enjoy.
There is no one here to argue with, no one to break
the serenity. Take a day or two away from your
cares. That's what we agreed upon. Don't slide back
on our intent now. For my sake, if not for your own.
Remember what we said. Revitalization."

He looked at me over the table of soiled plates
and empty cups. "You are an intelligent man, a good
friend, and such a fool. You know what preoccupies
me. That man. He's either here to do me harm or
protect me. If you didn't hire him, it's the former." I
started to protest, but he held up his hand. "No white
lies. I hope it's the latter. If it's not, and you're not
worried, then I won't worry either. Let us just enjoy

the day."

It was such an odd thing. I had been hiring bodyguards for him and lying about it. As far as I knew, the odd stranger was just that – an odd stranger. I was in a position wherein I couldn't say anything that would ease Galileo's mind. He probably wouldn't believe the truth, that I knew nothing about the man. Or, if he did, he would have real cause for worry. A lie, for instance, that I hired the man, would give us both cause for worry, because Galileo could confirm that I had been lying for years and I would start wondering about the man on the horse and have the same concerns as my friend. I really needed to not think about it, and Galileo had just given me that out. "Then let's just get on with the hike," I said.

***

It was a glorious day. The mountains had forced a few clouds to form, but they were lazy children with no intention of becoming adults. They would stay little puffs of white all day, forming and evaporating over the peaks. The old Baronet's castle, overlooking the commune from a rise at the end of the valley, was a benign presence. My expected peace was settling on me, and my friend was even smiling and breathing easily. He had acclimated to the altitude overnight, and seemed in better physical condition than I had anticipated. We could climb at a reasonably good pace.

I led my friend to the head of a favorite trail. I knew every switchback, almost every rock. It was for me like being in the embrace of an old comrade. I tried to read its effects on Galileo, but it was difficult to pull away from my own sense of pleasure, so I could only note an occasional grunt of approval or a nod at a particularly interesting bug. An hour's hike put us at an overlook, a single rock jutting into space, a place that allowed a view of the entire valley, at an elevation essentially equal to that of the Castel Tirolo. When our breathing evened, the only noises were the birds, the slight rustling of wind in trees, and if one listened closely, the murmur of a small brook close by. We sat wordlessly. I drew from my vest a paper containing a poem I had written the first time I visited this very spot. I had never shown it to Galileo, and I had no intention of showing him now. I read it again to myself. It was my mantra.

I'm through with you, Yaweh
I've tried
But for all your whines and threats
You have never delivered
Kill those Philistines, you cried
Kill those Canaanites, you exhorted
Cheer on the Seven Plagues, you commanded
And all for what
To remediate Adam's mistakes by turning it over to
Noah
To remediate Noah's mistakes by turning it over to
Jesus

To make Jesus pick up after himself by coming a
second time
You ought to be able to do better
I know I can
I will just be good
I don't need Your finger in the pie

Galileo grew curious at my mumbling. "Are
you saying Matins? Why do you need notes?"

I was tempted to give an honest reply, even
show him the poem. But no, I couldn't. "Just
reminding myself of the names of some of the
wildflowers," I said. I couldn't expose my friend to
my heresy. He was too devout and would never
understand what I meant. He might understand my
thesis that goodness came from within, but no one in
Catholic Italy would accept that conceit. He would
be saddled with the truth about my apostasy, a
condition which he probably already suspected but
chose to ignore. There are things between friends
better left unsaid, so I kept it to myself. There were
other things I could share with him, however. "If
you're feeling refreshed, we can continue on. Up the
trail further we'll come to a small stream, right at a
beautiful waterfall. If your balance is up to it, we can
cross the stream just over the fall. A log has been
felled across it to act as a bridge. It requires a little
courage as well as balance and care, but it is quite
exhilarating. Do you want to try it? I guarantee it is
one of the most beautiful places in Italy."

"Of course I want to go there. Waterfalls intrigue me. I know that someday there will be mathematics to describe their motion, but that day is far in the future. In the meantime I can enjoy the motion of continuously falling objects, in this case the drops and cups and buckets of water, and ponder their interaction. It seems so chaotic, but it can be interpreted and understood on some level. I may not be able to tell you what I understand when I look at such things. Nor can you tell me, precisely, what thoughts you have, short of poetry. And you know I don't care much for your poetry. We can just accept that each has some sort of understanding about the nature of a waterfall. That is the wonder of the minds that God gave us."

I could think of no proper response, so I merely rose and motioned Galileo to follow me up the trail. Not far along, the trail became bounded by wildflowers, almost as if they had been purposely planted to beautify our path. I'm sure my friend would say they had, but not by an earthly hand. For my part, I knew that the trail collected water when the winter snows melted, so its borders were the perfect place for seeds to germinate. And there was a plethora of flowers, of many varieties. Gretabills, ladyhands, fulsometails, peterwicks, and others, the names of which escape me and, of course, weren't written on the paper as I had claimed when I hurriedly stuffed it back into my pocket.

Up we went, amid the deftly named flowers.

As I had promised, we came to the stream. To our right it ran down a slight cascade, already boiling and turning white in its desire to plunge through the gap and over the cliff. To our left, the rumble of the whitewater became a roar as it hurled itself into space and down to the rocks below. Also, as I had promised, was the log lying across it. The log seemed to be of smaller diameter than I had remembered, and certainly more moist than I had ever seen it. Had I actually crossed this faux bridge several times? I couldn't turn back now, since I had boasted of the spot so avidly. I tried to display complete composure and confidence, although I felt neither. "I'll go across first. Just to be sure it's safe." I don't know how my voice sounded, but I detected a look of uncertainty on my friend's face. I ignored it and stepped boldly onto the log. It felt even smaller than it looked, but perhaps not as slippery. With care and with certainty I trod, foot after foot, across the roaring water, ever aware of the void into which the torrent disappeared to my left. I was just more than halfway when I heard Galileo's cry, louder than the rumble of the water. Much against my better judgment, I turned to see what caused this. There was a man, a large, menacing man, standing over my friend. He was brandishing a very large dagger, or a very short, slender sword. I have never resolved what it should be called. I care not, and I cared even less at the time. "What.......?" I cried.

"I have come to kill you, sir" the man yelled. "And since your friend here is a witness, I shall have to do him in also. I think you have put yourself in a position to make that simple for me."

I started to protest, to exclaim my incredulity at this turn of events, and this caused me to totter. I almost fell into the boiling waters right then. Distracted, the man chuckled, and at that moment Galileo, displaying a remarkable quickness of thought I had never before observed in him, pushed himself back from the man, and threw himself onto the log, straddling it. He then inched his way closer to me. The would-be assassin stopped chuckling. "That's it, then? You want to make it that easy for me? Two birds with one stone?" With that, he reached for log, with the obvious intent of twisting it to force us both to fall. He was ostensibly a very strong man, and undoubtedly thought he could lift it easily. It was, however, water logged, and it had two grown men to lever. It must also have been as slippery as it looked, since he only raised it a bit before it slipped from his grasp. The jagged end fell squarely on his left foot. He cursed and tried to reach it, raising the left foot and hopping around on the right. He was too close to the bank, though, and when he lost his balance the slippery ground caused his right leg to slide out from under him. His back hit a rock at about kidney level. He let out a cry rather than another curse, and slipped into the stream. He was now clutching for anything that might be attached to earth and not water, but only succeeded in dragging his fingers like useless claws through the mud as the stream began to swallow him. I watched all this happen, as if in a slowed, dreamlike motion, all the while teetering myself, and trying to use the sitting Galileo for support. I consigned the man to Galileo's God, since I couldn't offer one myself. I was sure he was a goner.

It is strange how one's mind can produce incongruous, even extravagant thoughts at a moment of realized catastrophe. A dying man might think that he forgot to feed his dog that day. As I watched the man being swept over the falls, I recalled Galileo's speculations on the nature of falling water. Then, as the man snagged on the last rock above the falls, becoming pinned there by the pressure of the water, my mind wandered to the very nature of motion itself. I think I saw the world, be it ever so briefly, as my friend saw it. Things did stop, not naturally, but because something stopped them. If something didn't stop them, they would keep on going, forever. Or, to the bottom of a waterfall, whichever came first.

Such thoughts ameliorated whatever hateful ideas I might have harbored about the man. My racing mind was turning pages too fast for me to perceive. Could I watch that man die - that would be murderer - either by suffocating against that rock or being swept over the falls? Was I not a physician, dedicated not only to doing no harm, but to saving human life if at all possible? How many lives had I saved by tending to the pimples of the princesses of Padua? And what of my mantra, that poem I was afraid to read to Galileo for fear of being mocked? Did I not vow to be good, in spite of any god or gods? I realized, even as I feared for my own life, even as I fretted that my friend might not be able to hold on or make his way to either side of the log bridge, that I must attend to the man in deepest peril. I trembled as my feet moved, almost without my will, sliding along a path so fraught with danger that only the surest foot could expect to survive, and then I fell. Not into the water, but to the bank.

I should have fainted there, but something, I know not what, kept me conscious, thinking, struggling. It was just a short distance to the distressed man, just a few feet down toward the falls, and when I thought of that distance, I thought, strangely, of Lutheran men standing in a straight line, toe to heel. Thirteen of them. Why thirteen? The Last Supper? That thought passed quickly. The more immediate concern impressed itself on my mind. What was my plan? How could I reach the fellow, who so recently was my mortal enemy, and who now was the object of rescue? I noticed a large branch, which had broken from the very log that formed the bridge. As I crawled toward the pinned man, who looked as if he couldn't last much longer, I grabbed the limb and dragged it along with me. When I drew parallel with him I extended it toward him. It was barely long enough to reach him but I was able to prod him gently with it. "Here," I yelled above the torrent. "Take hold! I'll pull you in!" He was barely alive. With drooping eyes he glanced at me and shook his head. I yelled again, louder. "You must! It's your only chance. Grab hold. Tight. I will pull with all my strength!" He shook his head again, then seemed, glazed eyes and all, to pull himself to a higher consciousness. With some newfound resolve he grasped the branch.

It was relatively easy to pull him sideways off the rock, but once he was in the current I found it took all my strength just to hold him there. It became a tug of war between me and the stream. I couldn't muster any more force, and felt myself weakening. Despair began to creep over me, and I felt, ever more strongly, that I was going to lose the battle. And then, just as I sensed the last bit of energy drain, a pair of arms grabbed me from behind. The hands clasped in the middle of my chest, and a strong pull now aided me in my battle. Galileo had crossed the bridge, and joined in the rescue of the man who, moments before, had sought to kill us both.

Chapter Eight

*Continuation of excerpts from the memoirs of
Salvatore Omar Fishbeingnola, unspecified date,
year 1599.*

We had made a small fire to help dry the shivering would-be assassin. We didn't question him straight away, since he needed to regain a normal body temperature after his icy soaking and near death at the edge of the waterfall. Galileo and I were patient; we had the calm satisfaction of just having done the right thing. Moreover, our attacker wasn't in a hurry to leave the fire, nor was he in any condition to fight with two healthy men. He had neither the means nor the constitution to do us harm. Finally, he stirred and grunted, as if signaling a desire to speak.

"What's your name?" was my first question.

"That can't possibly have any meaning to you, sir."

"That's true. But I take it you know who we are. Or do you just go about murdering strangers at whim?"

"No, sir. I was paid to do you in. I have studied your habits for some time. The purpose of that was about to become realized here today."

"I knew I had seen you before," Galileo exclaimed.

"I must have been careless when following Signore Fishbeingnola. I was sure he never felt my presence. I didn't think about you."

"And how long have you 'studied my habits?'" I asked.

"Off and on for several years, sir."

I was astounded. "You have been planning my assassination for that long? Impossible. If that was your intention, why wait this long? More importantly, who paid you and waited patiently while you 'studied me?'" I was pretty sure I knew the answer to that last question.

The fellow, barely able to talk when we pulled him from waterfall, was now a casual conversationalist. Almost garrulous. "People come to me with evil intentions but without the courage to carry them out themselves. They hire me to do so. A man with a serious grudge against you hired me. I followed you, waiting for the right place and the right circumstances. He wanted it to happen unnoticed if possible. You could either have an accident while on a trip like this or just disappear. I felt if I waited long enough you would present me with the opportunity to make one or the other happen. You did that for me today when you tried to cross the log. You were then separated from your friend and both vulnerable. I had meant to dispatch Signore Galileo first, and then tumble you off the log finish my job. Things didn't work out as I figured. And now, I owe you my life."

"You know this is Galileo Galilei?"

"Who doesn't? He was to be a bonus. My employer hates him almost as much as he hates you." The man became a bit more reflective. "When I think about it, I shouldn't have taken the task." He turned to Galileo. "It happens I am fascinated by your writings. I don't know what you two did to fill my employer with murderous intent, but I do know you are better men than he. And, as I said, I now owe you my life."

I needed to know if I was correct in my guess as to the fellow's employer. "You might be able to repay us if you answer a few simple questions. And perhaps make a vow."

Galileo broke in. "I'm not sure I want to trust this fellow. Shouldn't we bind him as a prisoner and take him down the mountain? Let the prefect at Commune del Tirolo deal with him."

"I'd much rather know the 'whys' and 'wherefores' of this affair." I turned to the unnamed man. "You say you were paid. Who paid you? Although I think I know the answer."

"I'm sure you do, sir. It was Maffeo

Bellarmino."

"I was afraid of that. Can it be that he has nursed a grudge all these years? I had forgotten such a grudge existed, it's been so long since I've felt threatened."

"Sir, I can tell you, since he hired me personally, that the grudge will never leave. I could see the hate in his eyes. His servants tell me that every time he loses at love or at debating, or even in a drinking bout, he curses Signore Galileo, and then you. Especially you."

I didn't want to believe this, but I knew it had to be true. "Still, my man, other questions remain. How did you know to find us here? Have you been following me or my friend for years, then?"

"No sir. I learned of your recreation pleasures from men who had worked for you in Pisa, and one who had worked in Padua and kept track of you."

"What men?"

"Why, sir, the very men you hired to protect Signore Galileo. They are brethren of mine, in the same trade, influencing life or death for money."

It was Galileo's turn to express his confirmed suspicions. "So, Salvatore. You swore you had nothing to do with the beatings given to Maffeo when he tried to attack me. You professed innocence so sincerely I wanted to believe you. But I knew. I knew."

I started to explain, but our captive came to my defense. "If he hadn't hired those bodyguards, Signore, Maffeo Bellarmino would have killed you. Or had you killed. You should thank your friend. And by not telling you, he not only saved your life, but gave you peace of mind. He paid heavily to guarantee your safety. He impressed upon my friends who, and you may believe this, would have taken less from Signore Bellarmino to kill you. You have a very good friend here, sir."

Galileo looked at me wordlessly. I shrugged. "That's the way it was, my friend."

Galileo shrugged back. "And the way it still is, apparently."

I turned back to our captive. "Perhaps not. Tell me, my man, can I extract a holy vow from you that you will quit this business, leaving it unfinished?"

"I can make such a vow, sir, but it is unnecessary. I have already taken Signore Bellarmino's money, for a job undone. I will lose my good name in my chosen profession, and there might even be a price on my own head. If you let me escape, I think I will go north. I have heard that the German Catholics are hiring mercenaries for the upcoming war with the Protestants. I don't know when it will start, but I may be paid to relax for a while. Again, I might be in the middle of something as soon as I get there. That's life for men like me. In any event, it's a way disappear. No, sir. A vow is unnecessary, but I'll take one if you wish."

"I'm a pragmatist and you make sense. My worry is that you won't be the last."

"I can tell you which of your former employees is the most trustworthy. If you hire him to be your eyes and ears in the underworld, you will have ample warning of any further attempts on your lives."

"That is probably the best I can hope for."

"But there's more, sir." He turned to Galileo. "Signore, threats to you don't arise only from Maffeo Bellarmino. In fact, if you hire the man I suggest, you will gain some protection . It is his brother, Roberto, who will be your public tormentor. I feel, I know, Roberto associates with people more sinister than even Maffeo."

Galileo considered, turning his immense intelligence to this new problem. Then he said to me, "We must talk about this later. First things first." He turned to our still unnamed attacker. "I don't wish to know your identity. I do need, for peace of mind, to have you tell my friend Salvatore how to contact the man who is to be our source of intelligence on the activities of Maffeo Bellarmino. And of this sinister force that dwells with Roberto."

"That is quite simple, sir." The assassin turned to me. "It is the man known to you as Amadeo. The one with a scar on his left ear, which came from a knife fight. You have used him for tasks, both legitimate and unsavory, in the past. If you let me escape Italy, I will talk to him before I leave. I will tell him of my debt to you. He is in the dark businesses, but he understands a debt of honor. If you pay him well, he will ensure the safety of you both. At least from Maffeo Bellarmino."

"And you, my man," said Galileo. "If we simply leave you now, can you make it down the mountain on your own? Your body heat has returned, but your foot seems damaged."

"I can take care of myself. If we part company now you will never see me again. I would appreciate it if you would tell no one of my intentions concerning the German Catholic Army. I will tell no others myself, not even Amadeo. I want to be rid of any association with Maffeo Bellarmino. To be truthful, I never liked the man. He paid well, and kept me on retainer so that I could live graciously for a long while. But he always treated me like a street thug, which I'm not."

"That is evident from your manner of speech and your literacy. Still, you have chosen an odious profession."

"So now I will become a mercenary. In your eyes, I suppose, not an improvement. Please then, simply forget about me."

"We, from this moment, know nothing of your whereabouts. For that matter, even of your existence."

"Thank you, sirs. I will advise Amadeo of your honor and my debt. Goodbye."

***

We were in Galileo's room at the little inn in Commune del Tirolo. We had some grappa made locally. It was quite good, having been flavored fully with anise seeds. It was also quite strong, so we drank slowly, a sip at a time. The room was warm and comfortable but otherwise sparse. It had a bed, a table, a porcelain wash bowl, and one chair. Galileo sat on the bed, I next to the table. My glass of grappa fit exactly into a ring burned into the top finish of the table, one of many made by spilled strong drinks of the past. My friend had been fulminating about the Bellarminos from the first sip. Now he let his fulminations slip into reflection. "Maffeo I can understand. There was the thing with the woman Bufana. And he insisted on intellectual arguments when he was in fact a ninny. He was destined to lose such confrontations.. Fools, and people without grace resort to violence when they lose contests which they should never have entered in the first place. So, I understand Maffeo's problems with me, and now, thanks to the near assassin, with you. Isn't it ironic that I would get the truth of this matter from a paid attacker, and not from my friend?"

I decided not to answer that and just took a sip of grappa.

"But, Roberto. Roberto? Why is he my antagonist? I have no quarrel with him spiritually. If I am a believer in the Copernican Doctrine, it is because I have done experiments that prove it to be not only possible, but rather more likely than the ideas of Ptolemy. This does nothing to deny the existence or the order of God. It does nothing to nullify the teachings of Jesus. It merely says that Aristotle, a human, was wrong. The church should be able to accept that, easily. It would not impinge on any of the arguments Bellarmino has made concerning true Catholic Doctrine. Does he think I'll undermine his arguments with the English Protestants? Does he know I've started communication with Kepler? Why should he care? It seems as if the Earth must be immobile else the very Church will crumble."

I took another sip of grappa and interrupted my friend. "It may not be about any truck you have with Protestants. Remember, Copernicus was a pillar of our church until he put the sun at the center of the universe. Even then, he didn't completely disengage from Ptolemy, if I have my history correct. Did he not keep the deferents and epicycles? Granted, he made the Earth move, but he made the center of the universe pure and perfect, in the sun. (Back to a Sun God, I said under my breath.) No, it's not about arguments with the Protestants, or even, as you seem to think, about the Earth being mobile. No, I believe it is about you personally. I believe he doesn't like you; perhaps even hates you more than does his brother, whose emotions stem from venal things. I'm convinced that Roberto Bellarmino thinks you are his superior spiritually. You do these physical things and yet realize they are connected to your God. You have brought the hand of God to reality, with mathematics and wood and metal and glass. Bellarmino suspects this, even if he doesn't quite understand it."

"Nonsense. I make toys to learn things. *Scientificus.* I suppose, since I teach myself these things, we could call what I do 'science'. But it is all within the framework of an infinite God. My science doesn't contradict Bellarmino's Doctrine. It merely helps it meet with the realities here on Earth."

"Ah, but don't you see? Bellarmino suspects that there are things other than earthly machines at work here." I pause for another sip. "Tell me something. Is it not true that you and I were born the same year in which Michelangelo died?"

Galileo smiled. "And is that supposed to mean something?"

"Have not some people suggested that the soul of a genius passes to another, preordained genius?"

"And you believe such drivel? Besides, why shouldn't you get his soul? Or, if you're too modest, why didn't that Englishman who puts on those infernally good plays get the soul? Why me? It's just silly to think such things. Souls are personal things. They travel with the owner."

"I don't know. People like you and Michelangelo come along once in a lifetime. Would your God throw away an extraordinary soul? Or lock it up forever with a dead man? Why not recycle it?"

"So why didn't Kepler get Michelangelo's soul?"

"He's German. The way I see it, the soul can go only to an Italian. Or maybe an Englishman."

"Why an Englishman? Why not a German?"

"It's not clear to me that Germans have souls. And just maybe you're right, and Michelangelo's soul went to the playwright in England. I can't be absolutely sure about these things."

It was Galileo's turn to sip his grappa. Instead, he took a large swallow, and finished off the glass. He looked at it, spun it slowly in his hand, then dropped it to watch it shatter. "If you're correct, I would like to be able to observe this world after my death. Just to see if some Englishman is able to explain things I can't, is able to make mathematics I haven't. I know that will happen. Sooner or later it will happen. Unlike you, I would allow it to happen to a German, or even a Jew like you. You all have souls, you know."

"Bellarmino would say that my soul is unredeemable. Apropos Bellarmino, I have some advice for you. You probably won't accept it, but I offer it anyway. Do your work. Do your science. But keep it to yourself. Don't wave it in his face. Publish only in Latin. Communicate only with others who do this 'science' thing of yours, and then only in Latin. And if they're Protestants, don't tell the Pope about it. Don't tell anyone who might tell the Pope."

"I love you like a brother, my friend. But my work is for the world. And if my soul is only for Italians, then Italians must learn of my work. In their own language."

We drank a lot more grappa that night. I wasn't then, nor was I ever, able to convince Galileo that he could neutralize the threat of Bellarmino by small compromises in his principles. His 'science' wouldn't let him compromise.

***

In the morning we started back for Padua, thinking we could return to the routine problems of our daily lives. Galileo's life was more problem filled than mine. In fact I was so happy with my personal life I considered my only problems those associated with Galileo. However, one of my very own was beginning to brew..

***

Notes for the memoirs of Salvatore Fishbeingnola, c. 1600

There was an advantage in having the clientele I did. It allowed me to associate Vito with the children of the best families. Then he would already be friendly with classmates when he matriculated at the best schools. Moreover, it would cement our Catholic credentials, since the Lantieris, the Fangios, the Bellotis, et al, were not only devout but prominent patrons in their parishes. Of course, so was I. Not devout, although I acted so, but a patron, and a generous one at that. I didn't pay much attention to the child of any particular family. I wanted Vito to pick his friends for himself. I knew sooner or later I would have to bring up the subject of Judaism, which would be our secret. I had to wait until he was old enough to understand what his background truly was, and why it was a secret.

Because I had grown to love my everywoman Anna so much, I had told her my background within a year of the start of our relationship. Her response had been, "So?" That non-problem had reinforced my resolve to continue passing as a Catholic, and to sooner or later give Vito his choice of religions, or to have none at all, but convince him that life in Italy was simpler if one at least pretended to be a Catholic. He needed to think about his place in the world first.

When one thinks of the Italian Renaissance the mind turns to Florence, or perhaps Venice, but the rebirth of the mind and soul, while emanating from those two cities, spilled out to the world through Padua. We Paduans were the funnel through which freedom of thought flowed north to the rest of Europe, and through which the reactions and thoughts of the rest of Europe flowed back into Italy. We were the ambassadors, and therefore we accumulated friends and correspondents in many countries. This gave me a safety valve for my rather heretical thoughts, which I could trade in conversation with certain people in England while remaining a hide bound Catholic in Padua.

As I have said, when my son Vito was too young to formulate complex thoughts on the questions of religion, morality, and spirituality, I didn't wish to abuse his mental development by imposing any beliefs on him. I let him take any religious training as a normal part of childhood, no more important than secular scholastics or the energetic play of young boys. I also let him prioritize his friendships himself.

It didn't take him long to decide on a best comrade. Federico Fangio was, even at an early age, a puzzle. He was kind to animals, well-behaved, and with a sense of humor born of curiosity. He would ask, with a smile, "Why are we surprised when ants appear to like our best cuisine? It tastes good to us. Shouldn't it taste good to them? Wouldn't they cook the very same thing if they had kitchens? Isn't that why they use ours?"

Vito and Federico didn't look like brothers. Vito had my green eyes and dark hair. Federico had very blue eyes and reddish hair like Galileo. I could see they would probably grow to about the same height, but I suspected Federico would be much more muscular. Vito seemed to take things as they came. Federico, from the start, wanted to change things. Federico had a thin nose and a broad face, while those two features were reversed on Vito. They seemed to love like brothers in the first few months of their friendship. This pleased me, since Vito would never have siblings. His mother was dead, and Anna, the only other woman I could ever love, wanted nothing to do with motherhood. For that matter I wanted only the fatherhood I already had. So, surreptitiously and without asking the Fangios, I appointed Federico a proxy brother to Vito. By the time I did this I think Vito had already done the same thing. For me this meant that there was one other being in the world for whose safety and security I would be responsible. I told no one but Anna about this. I told her everything since she had become the second half of my soul. She knew it was to be kept our secret, since it would be unconscionable to involve an innocent child in the dangers that accompanied my friendship with Galileo Galilei.

Unbound leaf inserted into the memoirs of Salvatore Fishbeingnola, date unknown.

Upon further thought, I could identify other enemies of Galileo, although they were more contests of egos and beliefs than of mindless hatred. Two came to mind.

Christoph Scheiner. He was a Jesuit theologian and astronomer who had a long running dispute with Galileo about *prima genitor* concerning sunspots. Many astronomers had noticed sunspots, most trying to tie them to transits of Mercury, since actual spots on the sun would ruin Copernicus' argument that the 'perfect orb' was the likely center of the universe. Actual observation made transits of Mercury impossible. It is likely that others first mentioned sunspots in various papers, but Scheiner and Galileo gathered good data on them, and spent years accusing each other of plagiarism.

Orazio Grassi. Another Jesuit astronomer, his dispute with Galileo concerned the nature of comets. Grassi was a supporter of the Tychonian Universe, put forward with good mathematics, which was geocentric like the Ptolemaic system, with the Sun and stars orbiting Earth and the planets orbiting the sun. It was acceptable to the Church.

Not all Jesuits were in conflict with Galileo, but those who knew him exhorted him, as I did, to stop publishing in Italian. He should have listened to us.

Chapter Nine

If Galileo weren't so busy inventing
Science, if his fame weren't growing throughout
Europe, if his experiments didn't consistently indicate
that the Earth moved, that inertia was a better law of
motion than Aristotle's Natural Law, or that there was
more than spiritual force that made Nature abhor a
vacuum only to thirty two feet, he might have been a
brooding man. As it was, he led the parade into the
world of, for lack of a better word, truth; the truth of
how things really worked. Throughout the first
decade of the Seventeenth Century he felt the world
coming to him

As people outside Italy had their impact on
Galileo, he in turn had impact on them. Tycho Brahe,
after falling out with King Christian of Denmark,
moved to Prague to serve as astronomer to Rudolph
II, the Holy Roman Emperor. He was joined by
Johannes Kepler, who needed the job but couldn't
stand Tycho's licentious ways. Tycho was possessor
of the best astronomic instruments to date, and thus
built the best catalogue of celestial objects to date.
He was great at astronomy, as great as he was at
wine, women, and song. Kepler didn't have to put up
with the last three for very long; he inherited the
position of court astronomer within a year, thanks to
said wine, women, and song. Using Tycho's data on
the planets, he could verify, much to Galileo's
delight, a solar system similar to that proposed by

Copernicus. He just did it with real data and his own mathematical model.

It didn't hinder Galileo's research when Giordano Bruno was burned at the stake for heresy, connected to his public declamations on the Copernican doctrine. Since it was Roberto Belarmino who instigated the stake burning, perhaps the man now practicing a discipline that he called science should have taken more note. That man, Galileo, was more likely to heed people like Francis Bacon, who, in Advancement of Learning, denounced magic and encouraged the scientific method.

And the scientific method is what Galileo did. He determined conclusively that that the distance covered by a freely falling body increases with the square of the time of fall, and that the increase is the same for all bodies, regardless of size or weight. Thanks to the 'lens makers' he had exhorted to hurry up, he was presented with the telescope, which he immediately improved, and then turned into an economic enterprise. He supplemented his income by selling telescopes to Venetian ship chandlers. By reading the flags on vessels returning from the Orient, the chandlers could tell what cargo was aboard. The chandler who could read the flags first could turn the most profit in the futures market. A Galilean telescope became a necessary tool.

It was the pointing of these devices skyward that gave Galileo the proof he needed to make Copernicus correct about the sun and planets. At first he noted that the planets, unlike the fixed stars, presented as discs, and not sparkling points of light. He concluded that the stars were much farther away than the planets. With an improved telescope he was to find the moon not to be smooth, but covered with mountains and flat areas he called 'seas'. He found 'stars' around Jupiter, which, as shown by their motions, turned out to be moons. It led him to ask the biting question, aimed at people like Bellarmino, "If the Earth is the center of the universe, why does Jupiter have moons?" He saw the rings of Saturn, and couldn't decipher them. He noticed that Venus had phases like our moon, but the disc of Venus was larger when the planet was in its full phase, and measurably smaller when it was in its crescent. This could only happen if Venus circled the sun. This posed another snide question for Bellarmino. If the Earth were the center of the universe, why does Venus circle the Sun? On the disappointing side: Copernicus' perfect sun, the ideal for the center of the universe, turned out to have spots. They weren't stationary, but they were there. What could be happening on the sun? There was so much to do.

Excerpts from the memoirs of Salvatore Omar Fishbeingnola, c. 1611.

I thought, after the incident in the Tyrol, my friend would be more careful in his work, in his march toward a natural world that had no use for the supernatural in its daily workings. That wasn't the case. Not only did he persist in trying to make the Earth move, he got into arguments with everyone - fellow scientists or Jesuits - it didn't matter. I could never quite follow most of it, but I knew he was annoying people in high places.

There was something about Saturn. To prove that he saw whatever it was first, he wrote to Kepler in a Latin anagram, something about the planet being triplets or something. Although he did this before anyone else announced the same thing or something similar about that planet, the arguments arose. And there was the business of the sunspots. Others claimed to have observed them first. Of course, my friend had become so famous that he was going to get the credit. And the enmities.

Fortunately, my practice was rather undemanding, so I could attend to the various nefarious activities I had been assigned by what I called the Tyrol Agreement. I had contacted my old conspirator Amadeo immediately upon return from the mountains, and he had agreed, for what I considered a reasonable price, to be our informant on the plans of Maffeo Bellarmino. Maffeo went dormant for a while after the failure of his assassination attempt. He did, through his contacts in the underworld, conduct a search for the man he hired for the task. It would prove unfruitful, since he didn't think of the German Army, and he obviously couldn't ask the only people who knew the answer, Galileo and I. He did, however, begin to hatch further plans of attack on us, all of which were relayed by Amadeo.

If thwarting these plans weren't enough of an avocation, the social lives of Galileo and myself would have kept us busy. For my part, I had settled into a nice living arrangement with Anna Bolen, and this tempting Swede had kept my house tidy, my sexual appetites satisfied, and my son Vito Abraham growing toward manhood. All she asked was to be treated as an equal, and to be paid enough to make her old age comfortable. I would never tell her that treating a woman as an equal was for me a mode of life. I actually felt they were, on average, more than equal. Men rarely rise to equality with women. As for the pay, although my parents thought it too much, I considered that, too, a bargain. Oh, I did have to remind Anna occasionally that although I was professionally intimate with many women, she alone provided my true romance. This intimacy with women had advantages other than sexual in nature. I kept that notion to myself.

My friend Galileo, on the other hand, had a rather more complicated love life. He had for years maintained a liaison with one Marina Gamba, who bore him children without the benefit of marriage. His religion told him these children were 'illegitimate', so there were certain sanctions imposed on his relationship with them, and upon their future. I had implored him many times to marry Marina, and solve his problems the easy way, since he had no romantic interest in other women. He was married to Science, and he simply couldn't bring himself to divorce that very demanding figure. I was sure, and I told him so as plainly as I could, that this was one divorce the church would happily grant. In fact, most of the clergy would be glad if my friend relegated Science to the status of concubine and married the mother of his children. That wasn't going to happen. I'm sure he loved his children and wished the best for them. He didn't see that he wasn't demonstrating this.

The benefit of this arrangement was that the Bellarminos didn't associate these children with Galileo, and hence intended them no harm, at least no harm beyond that which overactive piety inflicts upon the laity, as in the case of Roberto, or an uncontrollable temper inflicts upon perceived enemies, as in the case of Maffeo. This didn't mean that either of them relaxed their enmity toward my friend. Early on, Maffeo was especially active in this cause, and Amadeo reported to me frequently concerning nefarious plots to perpetrate bodily damage to Galileo or myself. And that is where my profession crossed with my private life. If Galileo had enemies is high places within the church, I had friends in high places within the civil government. Virtually all my patients were the wives, mistresses, or daughters of influential men. Some of the women were quite influential in their own right. They came in handy. I'll give just one example.

Amadeo came to me with tidings of a plot by Maffeo Bellarmino to harm Galileo by making it appear that a burglar had surprised him in his workshop, and in attempting to merely subdue and bind him, had strangled him by accident. He had hired a professional thief and murderer to carry out this task, a man from the south of Italy and unknown in the region. I'm convinced he found this fellow through his brother, after the man confessed to a Jesuit priest in Calabria, in an incredible description of maliciousness that no one would produce fictitiously. The priest was so dumbfounded he broke the most sacred of rules and told colleagues, of course without identifying the man. After sufficient time for the gossip chain to link Maffeo Bellarmino to the priest, Maffeo went to visit him. For some reason unfathomable to the priest, Maffeo requested the identity of the man. The rest is history, as I write it.

The man in question needed some help from the Paduan underworld to accomplish his task, and hence became known to my hireling Amadeo. With Amadeo and the help of one of my patients, a scheme was created to prevent this fellow from performing his job, without alerting Maffeo we were on to him. My patient's husband was the proprietor of a factory in which household goods were produced. He employed half dozen tinkerers of various sorts, all making useful products, in a shop which resembled, to the casual eye, Galileo's workplace. Amadeo, over a large amount of wine and other potent beverages, convinced the malefactor that his target had moved, temporarily, to said factory to carry out his experiments. He even made this seem conspiratorial, and asked for remuneration for saving the fellow the trouble of going to the wrong place. He convinced that man that my friend would be alone, working late on a given night, and that there was a bonus in valuable things to steal. The man could keep the money Maffeo was paying him. Amadeo just wanted a cut of the loot. It seemed very authentic.

My patient convinced her husband, easily done because it was true, that his factory was to be violated on that night by a criminal from Calabria. When asked how she knew, she also told the truth. Her doctor had a servant who had heard this from friends. The servant was reliable. Her doctor said so.

On the given night one of the factory owner's workers was stationed at a bench, under a dim light. He had volunteered to be the decoy, and appeared to be knowledgeable about the telescope with which he worked. The rest of the factory was dark and quiet. Just before midnight the criminal crept to the door nearest the worker, surveyed the scene through the window next to the door, burst in, and rushed toward the man tinkering with the telescope. He showed no priority other than murder, since he immediately attempted to strangle his victim. At that moment, three men jumped from the shadows. They were officers placed there by one of the magistrates of Padua. I should tell you that the magistrate's wife was also one of my patients.

The officers wrestled with the man, who was quite strong and intent on denying capture. This was his misfortune. In their struggles, the officers struck him several times; more than several, many, and several were hardy blows inflicted by clubs. This proved fatal to the criminal. That was never my intention, but it was not an unexpected possible outcome. I have often thought about such incidents. Since this whole thing was my plan, carried out for me, was I responsible for the death of that man, that person, that human being? There would be similar occurrences, with similar outcomes, as long as Maffeo Bellarmino lived. Was I in part responsible for any fatalities? In this case, I decided it didn't matter. The man was dead, he was a known thief and murderer, and perhaps a lesson had been conveyed to Maffeo. I had probably done the right thing. Mercenaries came cheap, and their lives were worth nothing, at least not to Maffeo Bellarmino. I wondered about his brother.

There was other wondering to do. Just a few days after the incident I just described, Marina Gamba came to visit me. She, too, was one of my patients, and I had helped her through her birthing ordeals. Galileo wouldn't have it otherwise. Now, however, she came not as patient, but as a conversant. She was a pretty enough woman, reminding me of my first and only wife, Gioia, now long departed. She only resembled Gioia physically, though. Marina had a much more aggressive personality than Gioia. Here in my examining room, which I kept sparse and efficient, she didn't have the air of expectancy of a usual patient. Rather, she let her strong personality show. She asked Anna, who was always present at my examinations, to leave. Without amenities, she began with a question, almost an accusation. "What did you have to do with the death of an intruder at

Signore De Paoli's factory?"

I tried not to look as if I knew anything more than town gossip. "Why, Marina, why would you think I had anything to do with that?"

"I don't know. But Galileo let it slip before he could catch himself that he suspected it was about him. And that means it must be about the two of you. I know you know De Paoli's wife. And you know the Magistrate's wife. Now I want to know what's going on."

This was all very difficult for me. Since the Tyrol affair, I had promised Galileo that I would keep him informed of any of Maffeo's attempts to do him harm, and of any involvement on my part. As a friend, I owed him that honesty. He had also made me promise never to let Marina know about such nasty business. "Marina, I can tell you nothing more about that business than you know already. A man broke into De Paoli's factory, with the intent of theft. He was surprised by a workman working late. As luck would have it, at least for the workman, he had some friends who were waiting for him to finish. They saved him from harm, but had to dispatch the intruder in the process. That's all I know."

"That's all you know. And what could be so valuable in De Paoli's place that a thief could cart away in a small sack? And the thief didn't even have a small sack. Was he going to heft a tin roof across Padua, to sell to a poor peasant in San Vito? And wasn't it convenient that the 'friends' were Magistrate's officers, trained in protection of innocents." The last wasn't inflected as a question.

"I only know what I've heard, which is apparently not as much as you have. I should be asking you questions."

"You and that lover of mine. Lover, at least, when I can get his attention. Why would he let it slip that it might be about him, and then pretend he never said any such thing? This makes me crazy."

"I can assure you Galileo had nothing to do with the happenings there. And I am just as innocent. You know my friend. When his mind's on his work, he says things that make no sense. Just words. He was probably thinking of the wording for one of his blasphemous treatises, and you mentioned the affair at De Paoli's, and he said it was about him, not even knowing what he was saying."

"For decent men, you two lie with great composure. I see that I can't pry anymore from you on this subject. But I have another question."

"I'll try to be more helpful if I can."

She extracted a paper from a pocket in her blouse. "Galileo gave me this love poem the other day. Read it."

She studied my face as I read. It said:

Love Across an Icy Bridge

Your figure is there
Framed by converging rails
A silhouette against a gray sky

Your face is unseen
But I know it
I know the expression
A slight frown, a sadness

Your thoughts run the ramp
Across the icy bridge
They meet mine
Exactly at a pane of ice
They cannot mingle

Yet still we love
Deep as the water beneath the river ice
Still we love

Of course I recognized it, since I wrote it at Galileo's behest. I said, "It's lovely. I think it says that he loves you as much as he loves truth itself. It's a wonderful pledge."

"Yes, it is. And it doesn't sound like him. Oh, I believe he loves me. But not that much. And not in the kind of words he never otherwise uses."

"You bring out something special in him." I struggled to keep my expression sympathetic, with the sympathy of an onlooker in a fragile romance.

She scrutinized me. "You wrote that for him, didn't you?"

I didn't budge in expression or in words. "Why, Marina, how ungracious of you. You attribute Galileo's expression of love to another man."

She grunted. "I can see that I'll get nowhere with you. Or with him. You can both have it your way."

As she left, I thought, "Oh how I wish that we could."

Further excerpts from 1611

What can I say about the continuing search for peace of soul? Only that life doesn't seem to reveal that destination to most people. Most of us are too busy dealing with our daily problems, sometimes small, sometimes seemingly monstrous, to peer over the horizon of our lives and see the path which leads to that lush meadow with the clear brook and the elm leaves waving in the gentle breeze. I thought I could take Galileo to the Tyrol and escape the nasty quarrels with Scheiner and Grassi, the felonious attempts at personal destruction by Bellarmino, the constant need to defend, defend, defend. I was wrong.

Now, to add to all this, after a bit of bliss in Padua, Vito added another major distraction to my efforts to smooth the road of life. I say 'major' because it concerned Vito and his brother-like friend Federico Fangio. Federico was having a falling out with his father and it certainly wasn't a minor family squabble. Federico, it seemed, wanted to convert to Judaism.

***

This business with religion has always annoyed me. As far as I could tell, everyone has a religion and most people have a Religion. The lower case is what one really believes and has taken on as a personal code. It is demonstrated in how one acts in daily life, from the trivial to the grandiose. It is driven by a person's conscience and dictates what he is allowed to do and what he should do. The upper case is driven by a person's need for guidance, rules, and a sense of belonging. It establishes the 'us', and makes possible a certain morality in a conflict with the 'them.' Both, however, are personal, and should be kept that way. That is why I have never felt guilt in posing as a Catholic when I am really a Fishbeingnola. To discover whatever that is, one would have to follow me around for a year, taking copious notes, then let me know what the rules of my true religion are. Even then one might discover contradictions. I'm human. By the way, a religion doesn't need a god, although almost all the upper case religions have at least one. If one wished to characterize their god based on the vague concepts in the minds of the congregation, one would find that a monotheistic religion has as many gods as there are members of that religion.

***

Federico was a boy for whom everything came easily, perhaps too easily. His father was rich, he was good at academics without trying, he was a superior horseman at a young age, he took to fencing quite rapidly, and he was handsome. He had a sharp mouth in classroom debates, especially with the masters, so Federico Fangio became Fang when one of his classmates learned the English meaning of that word. Fang was the bane of the careless professor, and the envy of lesser skilled students. My Vito couldn't see this coming at the age of five, but as Fang's personality developed it only endeared him more to Vito. Vito was just as intelligent and just as athletic as Fang, if I do say so myself. Even more handsome, I might add.

At any rate, Federico had a broad and eclectic field of interests. Padua, as part of the Venetian state, was a cosmopolitan city, offering the intellectual spices created by artisans from throughout the Mediterranean and the East. The search for different expressions of culture led him to a small shop on Ipagna Street in what was loosely defined as the Jewish District. The shop dealt in hand crafted household items, created to be useful and what some might call severe in appearance. Federico saw a geometric beauty in the utilitarian designs. He once told me, on a rare occasion when we met after the diaspora with his family, that he could 'feel the creator's thoughts in the simple yet precise geometry of the design.' He also found beauty in the shop owner's daughter who sold him a piece of that craft. He name was Rachela, and yes, she was beautiful. She was straight from the Bible: black, curly hair, eyes that were dark, swallowing vortices, a skin of creamy tan, the lithe movements of an angel, the full-lipped smile of a harlot, expressed on the face of an innocent maiden. I only saw her once, but that was enough to make me understand why Fang did what he did.

I can't go into the entire drama as it unfolded, because I wasn't present at any of it. I can only relate the major events that led to the outcome. Fang began to visit the shop of Rachela's father frequently, and he didn't make a purchase of something he didn't need every time he went there. He finally, though furtive messages to Rachela, convinced her to meet him in vacant archways near the markets where he could talk with her honestly and, he hoped, convincingly. She came to think as much of him as he did of her and so decided to make the fateful bargain. She was willing to introduce him to her father, which for her was a terrifying proposition. He had to understand that she wouldn't go against her father in any way. She loved the man, and wouldn't dishonor him by being an in-obedient daughter. There was something about her mother being dead and Rachela being an only child, but Vito, who told me most of this, wasn't quite sure. There were things Fang wouldn't tell even Vito. I've always had the feeling that the family history there was as complicated as our own. My fantasy was that Rachela was a bastard child birthed in a convent by a rich Italian woman to avoid scandal. The nuns adopted her out immediately, perhaps to the real father, rather than subsuming the daughter of a Jew. The woman's husband, a Venetian sea captain, was away for a year and never knew of his wife's pregnancy. His wife's lover got a beautiful child, the wife kept her good name, and the captain got syphilis from a whore in Alexandria. I'm not normally that imaginative, but I like that story.

So Fang had the talk with Elie Moberger, a proud Jew who had not Italianized his name to avoid troubles. Elie was not a strident man, not a threatening man, but he was a true man who loved his daughter and found spiritual surcease in his religion. He saw immediately that Fang was not the devil, not come to possess Rachela's soul, at least not in a demonic way. In fact, Elie could tell that Fang was already in possession of part of Rachela's soul, and she of his. There could be no enmity here, only resolve. So it came down to a simple pronouncement after a complex discussion. Elie said, in a calm voice, "You may marry my daughter if you convert to Judaism. If you choose to undergo that process, I will be glad to have you as my son-in-law, I will help you through the difficulties and obstacles, and I will stand by you in the troubles that are certain to follow, both with the Gentiles and the Jews. I believe you are a good man, perhaps the best my daughter could ever find, and that is one of the things I want for her. If you refuse to convert, I'll understand, and perhaps Rachela will as well. It is a monstrous undertaking for everyone concerned. I feel some trepidation, but I open my arms to you."

To which Fang replied, "I have a religion. I've been calling it Catholicism and making people believe it is true by practicing the trappings of that Church. A conversion won't change my religion – only the trappings and what I call it. I will gladly become a Jew, just as surely a true believer as I am now a true Catholic." When Elie protested that he didn't quite understand, Fang added, "What I am saying is that by any standards I will be a good Jewish man."

As Vito told me, Elie asked to ponder this for a day, to which Fang agreed with joyful confidence. And before he left, Fang, who had great facility in languages, learned two long prayers which he recited in Hebrew, with an Andalusian accent.

Chapter Ten

Many in Rome would blame Henry VIII for English Protestantism, but the real worry to Popes like Clement VIII, Leo XI, and Paul V was that damned James Stuart, son of Mary, Queen of Scots, baptized Catholic, subsumed into the Scottish Church by guardians, married by John Knox himself to a Scandinavian Protestant, and probably bi-sexual. The Henrys of France could be dealt with, even forced into Catholicism, but James I of England lay smoldering across a channel. Like Henry VIII, he seemed untouchable. Even Guy Fawkes couldn't blow him up.

Romans, those in the Vatican anyway, seemed to get Galileo wrong as well. He was a spiritual man, not in spite of, but because of his scientific pursuits. He was beginning to see the Earth as part of a grander scheme than Bronze Age myth makers could ever imagine. He was beginning to feel connected to the universe. He tried to get these feelings across to the people around him - clergy and laity alike. He felt that when the ancients declared that their god was unknowable, they didn't know how right they were. More importantly, they didn't even know what unknowable meant. Man could, however, know a little bit of the infinite, if he only tried. It seemed to him that nobody wanted to try. The laity wanted only to know that there was enough sausage to eat and wine to drink. The clergy wanted only to know what the laity shouldn't do. Galileo wanted to tell everyone what there was beyond sausage and wine, and let them decide for themselves whether God works through Church Canons or natural laws.

It turned out for Fishbeingnola's friend that it was a very bad time to mess with ideas about how the world really turns. Trouble was building across Europe, Protestant against Catholic. Corresponding with a Protestant astrologer/astronomer like Kepler, who cast lots for the good Catholic Rudolph II but who proved more and more that Copernicus was closer to the truth than Ptolemy, more and more that the Earth moved, more and more that there were many wonders out there in the universe, far from Earth and not in crystal spheres, more and more that Galileo was not just right but brilliantly right, didn't sit well with any Catholic Church authority.

Excerpts from the memoirs of Salvatore Omar Fishbeingnola, c. 1616

Just when I thought Maffeo Bellarmino would never quit trying kill us, someone else solved that problem. Just when I thought that this meant Galileo's troubles were over, they were really just beginning. That's because I didn't pay much attention to the man who warned us of deeper troubles before he ran away to join the German Catholic Army, and because my dear friend would never listen to me.

First, Maffeo. By 1615 I was a very

prosperous doctor, with investments in other businesses, as suggested by my father. These were done on my behalf, through him and by him. He in turn had gone beyond making marbles to brokering the construction of country houses for the rich of Venice, Padua, Florence, and other northern cities. He wedded the best artists and architects with the richest clients, keeping both happy. Some may have suspected he was a Jew. However, he was, in spite of great wealth, not ostentatious but rather, unassuming. He and my mother had put together an elaborate, if secret, Bar Mitzvah for my son Vito, who was now slightly taller than I. For my part, I had put together, on a small piece of property just outside Padua, an expensive but tasteful shrine to my first and only wife, Gioia. This was much appreciated by her parents, who attended Vito even more than my own. My inamorata, Anna Bolen, was very touched by this gesture, and she periodically went to the shrine, which was in a peaceful glade and well-kept garden, to read and meditate. She knew I could never make such a shrine to her, so she shared it with Gioia, and had the benefit of enjoying it while she was still living.

Still, my prosperity, my pride in my son's progress and talents in all things, both worldly and spiritual, and my harmony with my nanny, housekeeper, and lover Anna, did not give me total peace of mind. Maffeo Bellarmino lurked constantly, at least in my mind, and was an active presence on at least a dozen occasions, with attempts to harm either Galileo or me. My faithful informant and operative,

Amadeo, kept me always one step ahead of Maffeo,
the frenetic Maffeo, so that we could fend off these
attempts, mostly by traps and other artifices, such as
the one previously described. It seemed Maffeo
would never quit drinking too much, never quit his
usually unsuccessful tries at womanizing, and
therefore never became anything more than a
frustrated, impotent, and extremely angry drunk.
When Amadeo offered, sincerely and with business
like detail, to do away with Maffeo at what one could
consider a reasonable price, I almost took him up on
it. If I weren't a doctor, and if I weren't sure that
Galileo would abdicate our friendship if he
discovered my part in such a conspiracy, I would
have paid Amadeo for such a service, and had done
with it. Could a civilized man who was friend to the
world's greatest scientist, lover of a beautiful woman,
father of a fine young man, Jew, Christian, and
atheist, commission murder? I hardly think so.

And then, in the summer of 1615, just when
the gardenias in my shrine to Gioia were at their most
fragrant, Maffeo Bellarmino himself took actions that
made my list of major concerns one item shorter.
Through my informant, Amadeo, I had received
periodic reports on acquaintances from our early days
in Pisa. Bufana Giacometti was of some interest to
me, since she was seminal in the troubles between
Galileo and Maffeo. Shortly after the incidents in
Pisa, Bufana had met a rough but steady fellow,
slowed down, eventually quit her life of promiscuity,
and married. They had moved to Venice and opened
a tavern, not a place grand in either appearance or

clientele. Bufana was barmaid and barkeep, but now slapped any hands that tried to grope her. She had put on some weight, but not so much that her nipples went blind. She was a proper wife and they managed a living, remaining relatively moral in a bawdy atmosphere. They cheated no one, and Bufana's husband was strong enough to handle most rowdies, having some loyal friends to help with those he couldn't handle alone. Coincidentally, one of these friends was the other man I hired, along with Amadeo, to watch after Galileo when Maffeo first made threats against him.

Here is the story Amadeo told me, as he heard it from a first-hand witness.

It started on a balmy night in Venice, although it was always blustery when Maffeo Bellarmino was around. The tavern of Bufana and her husband was not on a canal; it was far enough removed so that the smell of the small garden overpowered the smell of canal water. Some of the patrons sat outside, where Bufana served them while her husband tended the bar. She was in a good mood, laughing at the off color jokes her patrons made, but otherwise innocent of any flirtation. Then Maffeo staggered in. He had been to other taverns already that night, so the stagger was quite pronounced. His eyelids dropped, as did those of any drunk. At that time, however, it was a customary expression for Maffeo: drooping eyelids, slightly parted, expressionless lips, perhaps some drool. If he weren't drunk, one would take him for an idiot. As it was, one could take him for a drunken idiot.

He noticed Bufana immediately, not because he recognized her but because she was female. When she turned, a flicker of cognition crossed his face just briefly before he reassumed the idiot's mask. He sat, taking an empty chair at a table otherwise full. He thought of what he had just seen. Cognition became recognition. Yes, he remembered that woman, a memory filled with bitter thoughts and consequences. He waved her over. When she approached, he grabbed her left buttock with his right hand and pulled her close to his face. "Ah, Bufana, isn't it? Remember me? Our little trysts in Pisa?"

She tried to pull away. She certainly remembered him, but she remembered no 'trysts', since there hadn't been any. She had found him to be obnoxious from the start, and she hoped her husband didn't hear Maffeo's remark. "I know you," she said. "But I don't remember you fondly. If you wish to buy wine or grappa, I'll serve you. Otherwise, leave me alone and get out of here."

He wiped his sleeve across his face. "Oh, the little whore is coy. You weren't so coy in the old days." He tightened his grip on her butt. "What do you have for me, besides wine?" His left hand moved to her right breast. "Ah. Same old Bufana, I see."

She yelled. It might have been "Let me alone," but it was drowned by other people beginning to yell also. Bufana's husband heard the ruckus, and came running. He arrived outside just in time to see Maffeo tear Bufana's blouse from her right shoulder, exposing her breast, which he then tried to grab. Her husband ran to the entangled pair and struck Maffeo solidly with his clenched fist, the blow landing in the area of Maffeo's right ear. Maffeo turned to face Bufana's husband. Now his look was that of a surprised and confused idiot. The husband struck him again, since he was still holding tight to Bufana's buttock. This hard punch struck him squarely in the forehead.

Bufana's husband yelled, "Ouch!". He had hit Maffeo's head so hard he hurt his hand. Maffeo fell to the ground, twitched a few times, and lay still.

When Maffeo was beaten senseless by hired thugs in my employ, it wasn't the first time he had been knocked unconscious in a street fight. Nor was it the last. A man with a lifestyle and personality like Maffeo Bellarmino had many serious scuffles in his lifetime. It is my reckoning that there had been much damage to his head through the years which had weakened the vascular system. It is likely that the two blows delivered by Bufana's husband were entirely sufficient to cause brain failure. I don't pretend to know much about that part of the human body. I don't think anyone does. I'm fairly certain that I'm right.

In any case, Maffeo was dead, killed while he was in a fit of drunken irrationality, by a husband protecting his wife from harm. It would seem a justifiable defense of a loved one, and there was no intention of murder, obviously. All the witnesses to this scene were sympathetic toward the defending husband; some of them had a personal animosity toward Maffeo, and there were none, none, who would express outright affection for him. The husband had done the right thing. The outcome was accidental, perhaps appropriate. Still, everyone who knew Maffeo knew that his brother could make great trouble for Bufana and her spouse. It was decided, unanimously, to make the death a mystery.

Maffeo's body was found in a canal, far from the tavern owned by Bufana and her husband. I was given a clear picture of the discovery by my bodyguard Amadeo. The corpse floated face down in the dark water for several days before an old crone forced her backward son to roll it onto a walk. Then it lay unidentified for hours. Finally the authorities retrieved it and eventually notified Roberto. The now very powerful clergyman arranged a high mass and a grand funeral for what had been the ragged corpse of a degenerate. Thanks to that, I swore I would never judge a dead man's worth by the pomp of his funeral. What struck me as most amazing though was that such a plethora of accomplices and witnesses could keep their secret, especially in light of the power of the Bellarmino family. Somehow they did. I imagine, as some confederates of Amadeo confided in him, and he in turn confided in me, many people knew the true story. On the other hand, there were no wounds from any weapons on Maffeo's corpse, and the bruises on his head could easily have happened from a tumble down stone steps into the canal. He was a known drunkard, and was often found fallen down in a stupor. The canal where he was found was near one of his favorite nocturnal haunts, and that's where the investigation centered. After a few months, authorities lost interest in the case, and even Roberto Bellarmino had to admit, given his brother's proclivities for overindulgence in spirits, that the death was accidental.

I am ashamed to admit that when Amadeo told me of Maffeo's demise I felt a certain sense of relief, which is no shame considering the menace he produced, and also a sense of satisfaction, which is shameful. It is true that I once thought of killing him myself, and then was tempted by Amadeo to have him killed. I couldn't bring myself to either of those actions then, and I couldn't now. There was, though, a passing glee at the thought of justice in the death of Maffeo Bellarmino.

If, however, I thought that Galileo was now trouble free, I was mistaken. My friend was still experimenting and publishing. As long as he kept to things like extending Archimedes' work on floatation into what he called hydrostatics, he didn't run afoul of Bellarmino. He wouldn't do that, however. And, he insisted on communication with Kepler, who wrote things that invariably made the Vatican's Prohibited List. The death of Maffeo did not mellow Roberto toward my friend, and it took less than a year for that animus to manifest itself.

That was just the beginning.

More excerpts from 1616

That was just the beginning of official troubles for Galileo, and the middle of the continuing drama for my son Vito's friend, Federico Fangio. Fang.

When Federico went ahead with his conversion to Judaism he lost a father and gained one. The estranged father didn't merely disown and disinherit his son, however. He vowed to make life miserable for him at every turn, every encounter. His dreams of his son and prodigy becoming a great man of letters and marrying into a noble family had been dashed on the Rock of Ages, as if the Jews were recrucifying Jesus. That it wasn't the Jews who had crucified Jesus in the first place didn't matter. They were at it again, tempting with some Eve this young man who was even more important to Luigi Fangio than the original Jesus. Rachela became a perpetrator of the Original Sin all over again. Yes, Luigi saw Fang as Jesus and Rachela as Eve, which shows he might have had his Bible all mixed up but he was sure about the inherent evil in the temptress and the inherent martyrdom in the naïve man. If Federico was to be a martyr, so be it. He would be a proper one. That would disgrace the family less.

I couldn't understand a father disowning his son over religious issues, but that lack of understanding could be attributed to my apathy toward Religion in general. I couldn't let a promising lad like Federico be cast onto the streets, so it was that he came to live with us for a bit, until he could sort out his life and have some security while doing so. We treated him like one of our family while he fretted and grieved at the loss of his own. Since Vito was already like a brother it was easy for me to act like a father. Anna was always the gentle presence of a mother. That just came naturally to her.

Still, the day came when Federico said to me. "Signore Fishbeingnola, although I am overwhelmed by the love and kindness I feel in this house, it is time for me to stand on my own. I have prepared for the ceremony of conversion to Judaism, and Rachela will soon be my wife. I have started to make a living at glass sculpture, and I believe I can support us. I feel so very lucky – I wasn't without a father for long. You'll never know how important that was to me. I can never repay you for that. Thank you."

But Fang had already acquired a second loving father who had nursed him through the conversion process, all the while knowing that Fang might not believe in God at all. Elie Moberger considered Fang a blank slate and was convinced that faith would fill it with love and understanding. He also provided Fang with a place in the artisan community. Elie and I helped him establish a shop making the most exquisite glass objetes d'art in all of Northern Italy. He did this under an alias and sold the works in Elie's shop.

Luigi Fangio turned to the underworld to make trouble for his son, and Vito turned to me. After Vito had reached maturity I made him privy to my machinations with shady characters to protect my friend Galileo. Anna, my dear Anna, had thought it wise to tell Vito of my clandestine dealings with some of these less desirable denizens of the alleys. She pointed out that I wouldn't necessarily outlive my friend and that Galileo would never change, so that an apprentice guardian would be appropriate. I appointed Vito as my logical successor and introduced him officially to Amadeo, my passport to the underworld. Vito had seen Amadeo on several occasions and had deduced an important relationship. When I gave him account of the several times Amadeo and friends had been useful in keeping Galileo from harm, he gave me a teary-eyed hug, one of the greatest displays of emotion I ever saw from him. "I love you father. You are a good man, a good friend, and a good father," he had said. It was not long after that that he came to me for help. He wanted to do for Fang what I had done for Galileo. I agreed to let him make a formal business arrangement with Amadeo, and set my conditions. The prime condition was that I wanted to know nothing of arrangements and plans. It was to be between him and Amadeo. If Fang were to know nothing, so was I. Amadeo was delighted to have a new, rich client, one who, like me, could make him feel that he did some good in the world after all. He readily agreed that each client would have confidential dealings, shielded from each other's business. Bodyguard is, after all, an honorable profession.

At this point I stopped asking about Federico Fangio, and Vito volunteered nothing. As far I knew, Fang was at least alive and well, no news being what it is. Vito was becoming a good doctor.

Chapter Eleven

Excerpts from correspondence between Cardinal Roberto Bellarmino and Father Francesco Montepulciano, Jesuit priest and clerk for Pope Paul V.

January 2, 1616 (B to M)

And on another matter, Francesco, as you edit my writings, I want you to be sure that the name of Galileo Galilei is never seen, not even with allusions to the man. His fame, rather his notoriety, has grown too much as it is. I firmly believe him to be a heretic. As such, I must find a way to silence him. I've had enough troubles with the Stuart king and that fallen Catholic, the Holy non-Roman Rudolph. I can do nothing about the Keplers of the world, but I must find a way to seal the lips of Galilei.

January 11, 1616   (M to B)

Of course, your Excellency, on the matter of Professor Galileo, I shall make sure that your writings give him no prominence. Should I also reaffirm your concerns in the ear of his Holiness?

January 15, 1616   (B to M)

On the third matter, Francesco, say nothing to the Holy Father. I've been rebuffed once, and I think the matter best be kept quiet. There is a season for all things. Can I expect your loyalty on this matter, no matter how long it may take?"

January 20, 1616   (M to B)

As regards Galileo Galilei, I will act as you desire, as long as you desire. Your constancy is mine.

Excerpts from the memoirs of Salvatore Omar Fishbeingnola, c. 1616.

I don't believe I have ever said that my friend was perfect. Perhaps that's why I loved him. He had a divine spirit and very human frailties. I had the latter but lacked the former, so I tried to touch this intellectual supremacy through him, all the while trying to counsel him through the shortcomings of his humanity. I considered myself an expert on shortcomings, so it was maddening that he hardly ever listened. I knew how to succumb to temptation, and how to avoid it. I knew, sometimes, what was right, and more importantly, when right must be done no matter how hard. I knew when there was little difference between right and wrong, so one might as well do what was easiest or what felt better. Was one more cup of wine right or wrong? Toss a coin. Was murder right or wrong? Suppose it would easily solve a problem. Don't toss a coin. You know the answer. What about the more complicated questions? What was right or wrong in love? What was right or wrong in paternity? What was right or wrong in the profound questions Galileo asked his Church, in the contradictions to deeply held faith, with the equally deeply felt connections to the physical reality of the universe? I tried to tell my friend that right in these last cases was personal survival, stoicism, continuing the creation of 'science' without shouting it from pages of a vulgar book. By vulgar, I mean a book written for the man in the street. The Egyptian High Priests had the right idea, I tried to explain. *We alone communicate with the gods, we alone understand the movement of the bodies across the sky. It is our secret, and we will tell you what you need to know. No more.* Galileo wouldn't listen to me.

"Those damn Jesuits are looking through telescopes," he said. "Some of them know the sun is center of things, that the Earth falls round the sun and rotates on an axis. But they keep it to themselves, because their Jesuit master, Bellarmino, says the Earth doesn't move."

"You walk a thin line, my friend. You let it be known that you don't believe in biblical literality, you don't believe in actual transubstantiation.

"Do you believe in those things?"

"No, but what I believe doesn't matter. It's what I say I believe that counts. And I say I believe in those things because that keeps me out of trouble."

"And that's fine for you. For a Jew to not be a Jew he has to reinvent himself or lie. I suppose it's even harder for an atheist who was a Jew to pretend to be a Christian. But you need to pretend to survive, and you seem to be able to live with that. Personally, I am driven to speak what I know to be true, and to call myths myths."

"The trouble is, you have other sins that your clerical enemies could bring up, if they chose. And these might give you trouble with the patrons who support you and even destroy your good relations, so far, with the papacy. You can't afford to lose the very important support you have earned through your work, most of which is innocuous enough not to bother anyone in the church."

"What sins have I that are so bad? I've been a good father. Ask anyone."

"Of three illegitimate children. Children which you haven't even tried to legitimize. That constitutes adultery, and I don't know what all in terms of other sins. Some conscientious priest could make a list, I'm sure."

"You know, at this late stage, I couldn't embarrass my patrons by admitting to all those children publicly. In private, no one minds, and I deprive the children of nothing, at least in terms of parental care and love."

"You deprive your daughters of marriage through your reluctance to claim them and then have to provide dowry."

"They will make good nuns."

"Would you have made a good priest?"

"My daughters don't see the universe as I do. Their souls will be at peace."

"And your son?"

"When the time comes, I will have him legitimized."

"Somehow that seems unfair to me. But what do I know? I have only the one son, and he's legitimate in two religions and a Duchy."

"And that's why I keep you as a friend. You're good at what you do, as a doctor, as a father, and as a false Christian with Christian values. You are a nobler man than any I know. What I don't turn to you for is your advice."

"I'm flattered. I think. I would protest that my advice is sound. Think back. Would it not have been easier if you had married Marina at the first pregnancy?"

***

Actually, there were two Marinas. One, whom we shall call Marina of Venice, died in 1613. She was the real mother of Galileo's children. I know, because I was the attending doctor, and because, as I have previously revealed, because I was a friend of hers. I also know, because I had visited with her just weeks before, that she was in good health just before she died. Her death was listed, by another doctor, as 'natural causes." That could mean anything from dysentery to one of the fevers, a tumor to gagging on a pomegranate. She had loved Galileo and had sided with me in trying to quell his campaign for scientific explanations of everything palpable. She was also known to Maffeo Bellarmino, who was in her vicinity at the time of her death, and who had had dealings with the doctor who announced the cause. This made me think her death might not be 'natural', but her burial was swift and the point was moot.

The other Marina was married to a prominent man, had known Galileo for many years, and had accepted guardianship of his children when he was busy making trouble for the Aristotelians. She didn't love him like Marina of Venice did, but she was a fine foster mother, and this double Marina business confused everyone. In a way, it was social brilliance of the part of my friend, something of which I would never have thought him capable, for it was so openly done and yet so devious. One couldn't accuse him of having a bastard family if they didn't know for sure he was the father, or with which Marina, and the children were in the domicile of one, a married woman. If the trumpets play loudly enough, it doesn't matter if the king doesn't attend the parade.

All that was well and good, but my heart mourned Marina of Venice. I had begged my friend to marry her. I suppose he couldn't. He was wed to his cause from an early age. I knew his answer before he gave it.

"It might have been easier. That's of no matter. It was never to be. Less likely than I could have become a priest or a doctor. I knew, from the time we were schoolmates at Camaldolese what I would be, what I would have to do. I simply didn't admit it to myself until we were at medical school in Pisa. So, you see, no matter how much you try, you can't silence me, and no matter how much you might weep for Marina, I could never marry her. I will just keep doing what I'm doing."

"I haven't wept for Marina for a long time. And I'll never stop trying, not to silence you, but to muzzle you a bit. Speak in Latin to Kepler, or to the English intellectuals. Don't speak in Italian to Jesuit seminarians. At least not about the phases of Venus, whatever that means."

He laughed and clapped me on the back. In the meantime, someone else thought they had found a way to silence him.

Chapter Twelve

Pertinent extract of a Conversation between Cardinal Roberto Bellarmino and Galileo Galilei in the office of the Cardinal, behind closed doors. Later in 1616.

Bellarmino: These things I have mentioned are only reiterations of my stance on such subjects, all of which I have made clear in my writings. If another man, one whom I didn't know, one who wasn't educated and reflective as well as devout, espoused such beliefs concerning transubstantiation or biblical literality, I might consider bringing it before the council. A person on the street shouldn't have such thoughts. Since you have never tried to impose such thoughts on the laity, we see no need to suppress your reflections. We merely suggest you have come to incorrect conclusions.

Galileo: I always welcome discussion of ideas. It almost invariably enlightens.

Bellarmino: You have made that clear on many occasions. That leads now to the main reason I summoned you. There are some ideas that are not open to discussion.

Galileo: And you have made *that* clear on many occasions.

Bellarmino: Not as clear as I now intend to make it. The Sacred Congregation of the Index has condemned *De Revolutionibus Orbium Coelestium,* pending correction. It is therefore forbidden to teach or defend it. I warn you of this condemnation before it has been made public, so that you may anticipate the consequences. The Copernican Doctrine will almost certainly be considered heresy.

Galileo: How can the truth be heresy? Surely it is God who designed things the way they are. Why would it be heresy to describe God's design more completely?

Bellarmino: I thought it was understood: This is not open to discussion. If you are not willing to publicly deny Copernican Doctrine some unpleasantness might ensue. I'm sure I needn't go into detail.

Galileo: You realize that much of my work reaffirms Copernicus. More so all the time.

Bellarmino: And much of your work, this science as you call it, doesn't touch on this subject. That is the labor upon which you should concentrate. Teach mathematics, teach the mechanics of the air or water or temperature. Do not teach that the Earth moves. You are interfering with the heavens themselves.

Galileo: And I am not only forbidden to teach this subject, I am to denounce it as false?

Bellarmino: Yes.

Galileo: (After some hesitation.) Very well.

Bellarmino: One more thing.

Galileo: What more could there be?

Bellarmino: Just this. Stop enticing my priests, especially the Jesuits who occupy the Vatican, to peer through your telescopes. Pope Paul hears of this, and thinks you are proving Copernicus' flimsy hypothesis. You may show the young and impressionable your other 'sciences', in which I have some interest myself. This, however, cannot be reconciled with the scriptures, and priests who haven't had time to reflect on this might be swayed toward heresy. I can't speak the Pope's mind, but I must warn you. Be careful what you teach, what you espouse. And how you do it.

Galileo: I will be careful.

Excerpts from the memoirs of Salvatore Omar Fishbeingnola, late 1616.

When Galileo returned from Rome he was fuming. This is more than a metaphor. If Vesuvius belched flames as my friend did, Pompeii would again be trembling. "That mendacious scoundrel! He threatens me with official sanctions, sanctions to be made church doctrine, and then keeps it between him

and me. He wants me to publicly disclaim
Copernicus, without the public knowing about his
threats. As if I suddenly discovered that all the proofs
I have found were really not there, that it was an open
investigation, but unlikely to support Copernican
Doctrine. That I've been wrong, that Kepler should
be dismissed. And that some of his Jesuit priests
aren't secretly Copernicans. Oh, Salvatore, how can
this be? Scurrilous rumors started at the highest
levels of the church, and made to look like popular
chat. Can I take this? Can I submit?"

I patted my friend's shoulder. "What choice
do you have? Your patrons, rich and powerful as
they are, might not support you if it comes to a battle
with the papacy."

Galileo sat. He was calming himself,
becoming more like the man I knew, the man who
considered the conditions, considered the actions,
predicted the outcome. "I can say this. Bellarmino
was hesitant to invoke the Pope to his side. I had the
feeling that he'd rather the Holy Father wasn't
involved in this dispute. I believe that's why our
meeting was private, why Copernicus isn't denounced
ex cathedra. Bellarmino has great power, but not
absolute power."

"Great is great enough for a mathematics
professor and maker of gadgets. He can hurt you if
you don't desist. He can't touch the Protestants. As
you have told me, your correspondent Kepler has
made himself invulnerable in Bohemia by promising

to name the Tychonian star tables he is expanding the Rudolphine Tables. Not that I understand what that means."

This distracted my friend momentarily. "What does it mean? That Kepler has done monumental work on the subject, and will put together the most important catalogue of heavenly objects ever assembled. Moreover, he has hinted that he is working on mathematics that will not only prove, but improve, Copernicus' work." He started to lecture me further on this subject, but saw that wasn't what I wanted to, or even could, on a par with him, discuss. He had once begun to describe his curiosity as to why Copernicus kept the deferents and epicycles of Ptolemy, and immediately lost me. He returned to the discussion concerning the politic of the Church. With a sigh he said, "If I must, I will try to keep my writings in Latin, and mostly to Kepler and my colleagues in England and the Netherlands. I will not, however, deny my most promising students the advances in thought and understanding which we together uncover."

"You are pushing your famous rock across the ice on a pond that is melting. Do you think it wise to push against Rome?"

"You forget some things, Salvatore. Bellarmino is influential, but he isn't Rome. And Rome still has unsettled business with Venice, and Venice is a powerful state. Moreover, the Duke of

Tuscany has always been my friend, and does not fear the Pope as much as you do. Nor does he revere Bellarmino as much as the Jesuits do. I am not undefended."

"Bellarmino is still an Inquisitor. It was he who concurred in the burning of Giordano Bruno. For defending Copernicus. And he has become stronger since."

"For that reason, I will recant. Not because of Bruno, who was a Copernican for the wrong reasons. But I will recant in a very faint voice, so that those who matter can claim they never heard me. And I will continue to study. And I will continue to publish. And among my writings will be a description of the celestial bodies, their places and manner and motions, as I truly find them. I will place my ideas before the world, to let it decide what moons really mean, how the tides work, why things stop when it seems no force is applied. These things will be strewn among teachings on temperature and pressure and time. I will not relent."

I saw that I could not save my friend from himself. My hope was that the war that had finally begun in earnest in Germany would prove a distraction to the Vatican. The lines had been blurred for years, Christian against Christian, Catholic nobles with Protestant aides against Protestant nobles with Catholic aides. Pope Paul should be too busy to worry about a scientist.

Chapter Thirteen

Private conversation between Cardinal Roberto Bellarmino and Father Francesco Montepulciano, Society of Jesus, in the bed chamber of the former, at the end of the second week of September, 1621.

Montepulciano (M): Perhaps you're tired now, Excellency. This has been a long conversation.

Bellarmino (B): Yes, Francesco, but there is still the most important thing to discuss.

M: You refer to Galileo.

B: Exactly. I know that he has continued teaching Copernican Doctrine, and that he continues to communicate with others who also espouse that heresy. I have been gone from Rome these last years, and Professor Galileo has become bolder and more vociferous. While the new Pope Gregory studied as one of us, and may help hold Galileo in check, I fear for his health. It seems not much better than mine. And I don't know who the next Pope might be. It could be a friend of Galileo, a man as hardy as he is brilliant. You know that much of my work has contributed to the very foundation of the Mother Church and its doctrines, which are firm and literal, and irreconcilable with heliocentrism. When I am no longer able to influence the Papacy or the Index, they may let Galileo sway them. I recoil at the very thought of that.

M: You will always influence them, Excellency.

B: Perhaps not so much after my death. Not without someone to carry on my work for me.

M: I don't want to think of your death, Excellency.

B: However, you must. You must because I need you to be my surrogate in this matter, to not let things come to rest on the side of that scientist, as he calls himself.

M: You have my word. My oath.

B: It may not be easy, but you must be conscientious. (He closes his eyes, and his face relaxes, as if in deep thought.)

M: If I am to do this, and I will, how active may I be? Should I argue your cause, and argue only? (B. does not respond.) This is rather delicate, Excellency. There are....stories. Concerning your brother. And certain physical actions he attempted to...quiet Galileo and his local cohorts. (B. does not respond.) May I take such actions, abhorrent though they might be, in defense of the Church? Would physical violence before the sin is committed be justified,when we know the punishment for the sin is burning at the stake?"

Monterpulciano waited for the answer, but

Bellarmino was asleep. He died three days later, never having answered the last question.

Excerpts from the memoirs of Salvatore Omar Fishbeingnola, c. 1626.

Life is so unfair. Just as my son had become a man, in the study of medicine like me and in the study of business like my father, just as he was beginning to demonstrate the high principles I taught him concerning human life, dignity, and friendship, my best friend, Galileo, did it again. He forced me to prove my loyalty and friendship through another publication in Italian, this one called *The Assayer.* In it he presented some hypothesis about comets, which may or may not be correct. That wasn't the important part. The important part was that he portrayed the language of nature as mathematics; this was a way of saying that the language of God was mathematics, expressed through physical models. The physical models did what you expected them to do because you got the math right. If you didn't get the math right, they did something you didn't expect. Or, perhaps, didn't want. This was a very different view of God than the Most Holy Bellarmino held, and it created mumblings in the sanctums of several holy orders. Among the Jesuits it created some arguments that might have been troubling to the Most Holy Belarmino if he were still living. My friend was right. Some Jesuits were looking at the sky differently.

Yes, life is unfair. My helpmate, lover, surrogate mother to my son, the only woman with whom I could, in my maturity, have any intimacy, seemed to grow more beautiful with each passing year. Anna Bolen kept my house, was a most gracious hostess to friends and guests, and even tended the summer garden, the shrine, which I had created for Vito's mother and my first love, Gioia. I could have counted myself the luckiest man in the world, having all this and the friendship of greatest scientist alive, if he weren't such a damned troublemaker. As it was, his troubles came my way, as they often did. This time, they caused a great change in my life, and in the lives of my immediate family.

When *The Assayer* was published, a certain Father Francesco Montepulciano visited me. He was one of those inner circle priests, I think a clerk to the Sacred Congregation of the Index, or perhaps to one of the appointed Inquisitors. Probably both. I thought it curious that a priest from outside my Diocese, from Rome in fact, should call on me unbidden. I had posed as a Catholic for so many years I would have believed it myself, if only I believed in God in the first place. I had even taught my son Vito how to look like a Catholic, so that he did it better than I. Granted, he had fond memories of his secret Bar Mitzvah, but that was just a party to him, somewhat like certain feast days of saints which we selectively celebrated to show ourselves not only as good Catholics, but erudite ones. All this given, why would this Montepulciano fellow call on me?

"I'll come right to the point, Signore Fishbeingnola. You're a friend of Galileo Galilei?"

"Yes. I have that honor."

"And do you hold the same beliefs? That the Earth moves around the sun? That the Earth rotates upon itself?"

"To be truthful, Father, I understand little of that. My studies are of the human body. I understand a little more of that."

"And I understand that you are quite a good doctor. Your practice is attended by influential people."

"I see anyone who seeks my help, Father."

"And do these people hold to this Copernican thought?"

"I don't discuss such things with my patients. As I said, I probably couldn't hold an intelligent conversation on such subjects. I care only about their ills."

"Can you vouch that you don't hold to such teachings?"

"I believe what the Church teaches." I didn't

like this priest. I certainly didn't trust him. I didn't hesitate to lie to him concerning my religious beliefs, or lack thereof, since I had been lying my whole life anyway. As I have said before, there are certain lies which one can, in all good conscience, make.

Montepulciano looked at me, up and down, and even tried to chisel his way into my soul through my eyes. One of my qualities, perhaps not one to boast of, is my imperviousness to demons who tried to penetrate my soul through my eyes. I considered this man a demon. "One more question, Signore. Do you have any ties to the Jewish community? Friends? Relatives? Your father's background is somewhat hazy. Certain records have been lost. Others appear fabricated. One wonders."

I was outraged but forced myself to keep a calm appearance. "And because you know that I am a friend of Galileo's, you have investigated my father? I'm not sure I understand."

"We have found, from past experience, that some heresies are generated within the Jewish community. We need to know if people who hold them are friends or relatives of Jews."

"Father, I try to be friendly with all men, and I try to consider them my brothers. I won't be swayed from that view, and if you would make that into some sort of relationship with Jews, then I'm guilty."

"No, that is following the teachings of Christ.

It is, however, as a practical matter of maintaining order and piety, necessary to identify those who would deviate from the Word."

"Let me reiterate. I understand little of what Galileo does. I understand him better as a person and a friend. He is good in both cases."

"I will take you at your word. Let me just warn you. His ideas border on apostasy. If he were a Jew, and not such a good friend of Pope Urban, he would not still be a free man. I will leave you with that thought."

I let the man out my door with "I bid you good bye, good Father." In both the 'bye' and the 'Father' I didn't intend 'good' to be taken literally.

***

When I told my friend of this encounter, he merely nodded and sighed. "The man was a toady of Bellarmino's. He ruminates about the Vatican, trying to make trouble for me. My old friend, Maffeo Barberini, now Urban VIII, pays him little heed. Even his Jesuit brethren wonder about his motives and methods. You shouldn't worry yourself. He won't investigate you any further. I doubt there will be any trouble to you or your family from this man. Of course, I can't be bothered by him. I have too much to do, too much to say."

Galileo was right about many things, but not

about this. And he seemed to miss the irony: his most deadly enemy and his sturdiest shield from harm were both named Maffeo. One was dead; the other couldn't last forever.

***

So it was that I decided to make one last sojourn to the Tyrol with my son Vito, who had taken over the role of hiking companion from Galileo. There was a serious, life changing matter that we needed to discuss, and the best place would be in those mountains. I settled the plans with my son, sent a courier to Commune de Tirol to signal the innkeeper of our intentions, warned all my patients of our impending absence, packed for the trip, and spent a romantic evening with my love, Anna. Like most women, Anna possessed that extra sense, received through some kind of emotional antennae, and when we had satisfied our physical love and were tending toward sleep, she said, "Salvatore, there is more on your mind than this trip, or some hiking in the Tyrol. Something that has to do with Vito, and therefore with me. You know I love him as if he were my own son. I'm really the only mother he has known, and if you intend to do something that will affect him dramatically, you must tell me. You know that it will affect me also."

I had anticipated this moment. Vito notwithstanding, the matter also affected her, just as she said. I knew I would have to explain some things to her, sooner or later. I took her hand, but didn't turn toward her. "I feel the possibility of some complications in our life. Life is messy, no matter how much care one takes in trying to keep it neat. Emotions, loyalty, honor all bring pride and joy, but they have their own set of problems. I found great joy in my love for Gioia, and great grief when she died. You brought that joy back, and gave love to my child as well, doubling the love in my life. If there were only you and he and the tranquil paths in the Tyrol, I might be able to keep the messiness of life at bay. But there is my friendship with Galileo, a friendship of more than forty years. More than a friendship. It is my brush with greatness, a palpation of immortality. I could no more deny that than deny my own being. Which, by the way, started out as a Jew, as you know. Those two things, mixed in just the right way, promise the chaos of an explosion."

Anna, her hand on my cheek, made me face her. "Why is this troubling you now? As you say, you've been steadfast to Galileo since childhood. And you really haven't been a Jew since you escaped the Venice ghetto. If you ever were, really. Did you ever really have a religion?"

"I don't think so. My father wanted me to think I was a Jew, but he always covered the trappings of the religion. And the catechism of the monastery school never took. I learned the words of Catholicism, but never the feelings, never the sentiment. Certainly never the reasoning. No, the reasoning I got from my friend when he argued down the irrationalities of his religion. That's what I heard and added to my own reflections. No. I could find no religion I could accept, no god that made sense. And now, when I sense an ominous presence, it is in the form of a man, and there is no god who will change the intentions of this man. It is left to me to cope."

Anna, bless her, kissed me. "For us to cope. You know that I am with you in anything you do. Your troubles are mine. This is one of the reasons I never wished to take the wedding vows. They are words. You will know how I truly feel by my actions. So, then. What do you sense? And what can I do?"

"There is at least one priest, perhaps more, who carries on the thoughts of Roberto Bellarmino. Carries them even further, if the rumors I receive are correct. It isn't just Copernican Doctrine which is forbidden, but also Galileo's atomist ideas. And this may not be confined to the Jesuits. The Dominicans would like to prove that they are as pious and devout, as strong defenders of the faith as the Jesuits. In fact, some of the Jesuits, the so-called Scholastic Group, might secretly discuss the insights men like Galileo have made into the natural world. That doesn't matter. My friend is in Italy. The Scholastics are in Bohemia."

"But what does all this mean to us?"

I stroked her hair, the hair of this darling Swede, which had now become lighter and lighter blonde - graying, but still topping a beautiful woman.

"There might be some physical danger. Bellarmino's brother was always seeking thugs to do my friend and me harm. There might be a resurrection of such an instigator, in an attempt to somehow silence Galileo. And there is always the probing, probing into my heritage. To expose one of Galileo's friends as a rich, secret, successful Jew would tilt the debate against him. Worse if they discover that I'm really an atheist. It would convince the groups like the Index that his writings are tinged with the heresy of the anti-Christ. I don't know what's going to happen, but I want to protect Vito. And you. And my parents. There will be some kind of spill, I'm sure of that. From the chaos of unreasoned animosity, from the certainty of rectitude by placing words in the mouth of an incomprehensible god, there always is."

Anna returned the gentle strokes to my head, through hair that was thinning as well as turning gray. "You're too noble not to keep yourself in this drama. And you're too nimble to lose." Then she took my face in both her hands. "But, your friend Galileo....."

***

We packed well for the trip to Commune de Tirol. I insisted on packing many things for Vito, much more than needed for a few nights in a mountain inn. He was only mildly curious about this, and didn't argue. I assured him that I had good reasons. It might be that he wished to stay longer, or travel further. His medical practice was not yet so confining as to constrain him to short vacations. He should enjoy his youth more. I carried on like this for half an hour. He assured me that I needn't have. He wasn't bothered by a few extra bags in the carriage.

The roads immediately outside Padua were well traveled and well maintained - flat and smooth. Soon, though, we started up into the hills, and the bumps grew with the incline. Vito and I were silent, so I began, to the sporadic rhythm of the carriage, to repeat my new mantra to myself. I needed the reinforcement of this kind of meditation for what I had to do, for what I had to bring upon Vito. This meditation was the affirmation of faith, not in a god, but in men like Galileo, frail humans with faults, who discovered the course of the universe, bit by bit, mistake by mistake, triumph by triumph, but never by proclamation. The plan was no plan.

> *Omniyou say you're goodness*
> *Then you're not*
> *Omniyou say you're evil*
> *Then it's us*
>
> *You're omnidecided now*
> *So why believe your plan*
> *It doesn't make omni sense*
> *All this omni suspense*
> *It can be honestly said*
> *No plan is better*
> *A non-believer agrees*
> *Zero is an answer*
> *Omni isn't*

I said these last two lines over and over, because I had a plan for Vito, a plan that didn't include any gods. Events would soon prove that I should have included myself in my disclaimer on plans.

***

We were higher into the mountains, the road narrower. I thought the coach might be going a little fast. The coachman, newly in my employ, hadn't been over this road before, although I had described the route carefully, and warned him of the instability a large carriage might experience in some places. I was about to open the shutter to tell him to slow down, when, in fact, I heard him urge the horses to a faster pace. I tried to stand, just as we hit a large bump, and my head was thrown against the roof of the cab. Vito grabbed me as I fell back to my seat. "Let me handle this, father." He tried, through the port, to address the coachman, but the man merely looked at him with a kind of cold excitement; and then the face disappeared. We heard the horses whipped to go faster, then a figure hurtled past our door window. The coachman had abandoned the reins as well as the coach. And as we both realized we were driverless, the horses veered to avoid the steep incline on our right side; the carriage tilted, something snapped, and we tumbled over the edge, sliding down the mountain side, dragging two of the four horses with us. We would have continued all the way to the bottom of the ravine, but two sturdy trees, like giant arms, interrupted the slide. We came to rest, both of us on our sides, on the right door of our carriage. It took us a moment to gain speech.

Chapter Fourteen

Deposition by a witness to a carriage accident on the road to Commune de Tirol, Republic of Venice, August 27, 1626.

I make this account under oath to God and his son and my Lord, Jesus Christ.

On the road between San Pietro de Salmonella, the hamlet in which I make my Holy Obligations under Father Romolo, and Commune de Tirol, as I was taking one of my cows to pasture, I saw a very elaborate carriage leave the road and fall down the hillside. The carriage was traveling faster than it should on such a road, and I saw the coachman leap from the carriage just before it left the road. He struck the ground very hard, but was trying to stand as I reached him. I thought he was badly hurt, with blood showing on one arm and the other shoulder. I asked if anything was broken. He refused my help, and began to walk down the road without talking any more. I didn't see him again. I don't know where he went. I went to the edge of the road and saw the carriage on its side, more than one hundred feet down the hill, resting against some trees. The trees were at the top of a very steep slope that fell down to a rocky creek. The carriage would have been destroyed save for the trees. I went to the carriage and found two men still inside. Both had bruises and scrapes, but both had escaped any broken bones or punctures of a deadly sort. The older of the two, whom I helped out first, introduced himself simply as Salvatore. The second man was his son, Vito. I found they both could stand on their own. Salvatore told me that they had some few possessions strapped to the carriage, and some small bags inside. If I would be so kind as to fetch a cart to help them from the scene, he would reward me. I told him that as a good Catholic man it was my duty to help strangers in trouble, and would accept no reward, but it would take me more than an hour to return with the cart. He assured me they would be fine until then. I went to my house and fetched my horse and cart. When I returned to the scene, the two men were waiting with two large trunks and two large traveling bags. My cart was barely able to hold the men and their baggage, but we made it safely back to my farm. We washed their wounds and my wife prepared a dinner. Since they appeared to be high-class gentlemen and not accustomed to sleeping on hay, my wife and I decided to sleep in the barn, and give them our bed and house for the night. They seemed in good health by the next morning. It took two days to arrange transportation for them. Only one went to Venice. The other continued north, saying he was heading for Salzburg.

Alberto Schleigl, farmer and husbandman.

Excerpts from the memoirs of Salvatore Omar Fishbeingnola, 1627

I knew it wasn't an accident, and that the coachman hadn't simply fallen from the carriage. I also knew that we had been fortunate, very fortunate. The coachman had set us on a disastrous course before he jumped, one meant to end our lives, either by death or disabling injury if we lived. He had chosen the spot to ensure catastrophe. The horses, having sense not to run over the cliff, had veered in such a way as to put the coach on its side, so that it did not tumble but rather slid down the grade. Even at that, the two sturdy trees at the top of the steepest part of the declivity saved us. A few feet either way and we would surely have rolled all the way to the bottom, conveyance and men broken forever. The net effect of the crash, however, was to hasten the conversation I needed to have with Vito, one that would change his life forever in any case. Now I would have that conversation, not in a cozy room in the inn at Commune de Tirol, but in a rough farmhouse somewhere even more removed from polite society. Yet it occurred to me the event could also work somewhat in Vito's favor.

When we were settled for the evening, and the good husbandman and his wife were off in their barn, I bade Vito to sit in the only comfortable chair in the house. He deferred it to me, and took a three-legged stool in front of it, sensing I had something important to say. In fact, before I even began, he said, "Finally we get down to what this trip is all about. Did it take that accident to bring it out?"

I wasn't surprised. Vito is an intelligent and perceptive young man. It made it easier. "Obviously, it wasn't an accident," I began. When Vito tried to interrupt with a question, I held up my hand. "Hear me out. Completely. Then I'll answer questions." He settled on his stool, giving me his full attention.

"Because of my friendship with Galileo, certain factions have, I'm quite sure, decided that they could undermine his credibility by destroying mine. Failing that, they also believe they can frighten him by harming me or my family. These factions feel that their cause is righteous, and they believe in their core that their god's vengeance can therefore rightfully come at the hands of man - their hands, you see, because their god is the correct and ruling god, and anything done in his behalf is therefore good. There is no doubt that our coachman had been bribed to drive us to our death in that ravine; that he was not as unfamiliar with this course as he led me to believe. These factions have wealth, power, and a conscience hardened by what they consider to be moral rectitude.

"Since you were a child, even before you were born, they have been seeking to harm my friend Galileo. At first, it was for ostensibly different reasons - loss of face in public, or some romantic dispute. But as my friend's fame has grown, and it has become more evident the he is correct in his assessments of the physical world. As Venice has pulled away from the Papacy, as the Carolingians have had congress with Protestants, the more insecure these factions have felt. Insecurity breeds desperation. Desperation breeds extreme acts. Murder, or the ruination of another man, seems the best solution to problems the other man represents.

For these reasons I have felt my family to be in danger. I don't worry so much for myself. I must, however, protect those dear to me as best I can.

"Now, my father has passed, and my mother is near the end of a good life. There is not much anyone can do to disturb my progenitors. Your other grandparents are very devout people, far removed from Galileo. They are safe. That leaves me, Anna, and you. I've already said that I don't worry about myself. I have presented this predicament to Anna, and she won't, under any circumstances, leave my side. I believe she and I will be able to weather this storm. That leaves, in the end, just you."

As Vito started to become animated I said, "No, don't protest, yet. Let me finish. I know that you will exclaim you are as steadfast as Anna, that you are younger and stronger than I, that you will do battle with anyone who seeks to harm me, that you can take care of yourself. And, under ordinary circumstances, I would agree. I would even enlist you to aid me. But there are factors, other people, the kinds with which you have only recently associated. Against us are people of the lowest character, people of the shadows, people who do not fight with honor, people whose very souls are cheaply bought. Out friend Amadeo knows these people and how they work. That is why we enlisted him as a soldier in our own army of the Dark. We have trusted him for s very long time, and I don't lose that trust now, but he is the reminder and the key we need to understand the real world. Amadeo would do things for me that I could never do myself. His counterparts work for people less ethical than you or I, people who would condone actions like murder. Has Amadeo ever suggested murder as a possible solution to my problems? Yes. Have I ever accepted? No. Would our enemies accept such a suggestion? More that accept – they would make the suggestion.

"Now remember, it is me they're after. One way to destroy me is to kill you. One way to distract me from shielding Galileo is to give me great dilemma by shifting you into the threat of mortal danger that now lies over Galileo and me. I know you love me, and I know you are not a coward. But consider my conundrum. Lose my honor by denouncing Galileo, or lose my son to some murderer's hand. Denounce myself as a Jew to save my son's life, just to give him a life as an outcast, ready to be a victim of the next Inquisition? None of these things are acceptable. I will not be Abraham, which is the middle name I now wish you to abjure. If there were a god, and God himself told me to kill you, I would deny Him. If I would do that, then I would do anything to keep you alive and well.

"For these reasons I have planned, without asking you, a new direction for your life. I would never do such a thing, unless it was the only way to give you a chance for a happy life. Let me describe to you the scheme that I have put in place.

"First, there is hidden in the false bottom of my trunk a number of gold coins, sufficient to continue your travels from here to your final destination. More than sufficient. You should probably leave them there, drawing a few at time to pay your way. Never let on that you have any wealth in your company. This will help you avoid cutthroats and schemers. The money I pass on to you is from your grandfather, who asked me to surprise you one day with an unexpected inheritance. I don't think he had this kind of surprise in mind, but it turns out to be terribly appropriate, for he intended for you to use it in pursuit of some great goal in life. Let that goal be a happy survival, for my sake as well as yours."

Vito couldn't contain himself any longer. "Then you planned, even as we left Padua, some great travel for me? Did you expect this contingency? Where would you have me go? For how long?"

And now came the difficult part. "No, Vito, I didn't expect this violence. Not on this trip. Not this soon. I had hoped to explain my intentions on a quiet mountain path, in the privacy and serenity only Nature can offer. That wasn't to be." I drew a deep breath. "I would have you leave Italy, my son. For good. To make a new life where you are safe from any reprisals I might suffer, reprisals which would surely come down to you as well. I would have Anna go with you, but as I've said, she won't hear of it. You, however, must put my mind at ease. All I wish is your survival - your happy survival. If I am sure you're safe, I can deal with any enemies freely and without fear."

"That's not acceptable. I should be......"

I stopped him before he could complete that sentence. "Hear me out. It is your duty to carry part of me forward. It is your duty, as my son, to have a happy life if it is at all possible, because that is the only thing that will guarantee my own happiness in my old age. I ask you to make this sacrifice for me, and graciously accept this gift I give you." I was prepared for his frown. "Vito, you would be leaving no true love behind. You have had several casual romances, as I did when I was your age. You have not yet met the love of your life. You have only begun to practice medicine, and you would soon leave my side in that practice in any case. I want you to be more than a tender to rich women with cosmetic problems. I want you to be a good doctor. I will see that you have sufficient wealth so that you can render service to those who are truly needy, who can't afford that which they most need. Every man should have some sort of nobility. Mine is in my steadfastness to Galileo. Yours could be in the practice of that for which you have been trained. Medicine can be a noble profession. It hasn't been my particular contribution to mankind. As I say, I aid mankind in another way, although I doubt anyone will ever know that." I shook my head. I was straying from the topic. "I have planned very carefully for you to be a doctor in England. You may choose some other profession, some other interest. But it must be away from Italy, and a move to England would be the most facile."

"England? But I have only a beginner's fluency in the language, and a rather bad accent at that."

"That can be easily handled, and I've already taken such things into account. Here's how you can proceed. In London there is a very successful tailor, a Mister Francis Baker. His real name is Levin. Baker, and you must call him that, is expecting you. You will stay with him in London for a few months. There you will undergo intense study in the language, down to the accents and idiomatic usage of the upper class. When you are ready, you will move to Lincolnshire, to a house that I have purchased there. You will establish a practice there, as a doctor from London who has come to the countryside to tend to the exurbanites. You will make a practice and a name. And then, in a year or two, you will move back to London, established as an Englishman, a doctor, a socially active person, and a true value to society. You might by then even have a wife. Who knows?"

"And you made all these plans without even hinting to me what you were doing?"

"If I told you before I even knew if it were possible, you would have had very strong arguments against me. I probably would have wavered. I had to know the pieces were in place before I presented you with the idea. You can still refuse. You are a man, your own person. You should determine the direction of your life. I only wish that you choose it independent of your feelings for me. I only offer this as an aid. Go to Germany and be a soldier. I would hate that, but I would accept it. Go to England and join the tailor in his business. But go and do something, someplace where I can know that you will never be caught in any web of intrigue or violence that might befall me. I ask this for my sake, as well as yours. Face your own dangers, and defeat them. Just don't face those of my making. Those are for me. You must give me that." I put all the emotion I could muster into this plea. I was visibly sweating and on the verge of tears. I was glad he noticed this, for he needed to know how important this was to me. He needed to feel that this was the right thing for him to do. For me. For himself. I was counting on his love for me, his intelligence, and the adventuresome spirit of a young man, eager to go out into the world on his own. I will admit now that I was unsure of his response.

I needn't have been. He sighed, and looked at me directly, his eyes searching my face, my soul. He sighed again, and stood. "If I didn't know you," he started, "I don't think I could understand your motives. If I didn't know you, I would be uncertain of my own abilities to leave my home and start a new life in a strange land. If I didn't know you, I would doubt this as an act of love. But I do know you. And in my heart, I know this is the best thing, because you are so sure it is. One thing - this can't be goodbye forever. Can we somehow keep in touch, perhaps rejoin our lives sometime in the future?"

"Of course we'll keep in touch. Through the tailor Baker, and special couriers."

"How do you know this man? You seem to have placed great trust in him."

"Your grandfather was a man of international business, a man of great ethics and generosity. Baker was a good friend and sometimes agent. There was affection and loyalty both ways."

"Still, you have made intricate and expansive plans. Were you so sure that I would accept? What if I had said I want to go to Bohemia and do astronomy with this Kepler fellow?"

"Then I would have wished you well and given you the same stipends to see you on your way."

"And the house in Lincolnshire?"

"A gift to Baker. He has worked very hard on this, and kept very quiet. A remarkable friend."

"This is yet another lesson for me, father. And I thank you for it."

"What lesson? I intended none."

"But it's there. You've shown me that love and loyalty are a currency unto themselves. They buy people in ways that money can't. They influence people in ways that money can't. I have no doubts that your plan is the best course of action for me, and I shall do my best to make you glad you conceived it."

I felt a great relief. "But, you must, from now on, make your own decisions about how to lead your life. My idea has never been to control your existence. Just to protect it."

"I know." He looked around. "You don't suppose these humble people have grappa in the house. I would leave some of these gold coins of which you speak for something sturdy to drink."

***

Vito left me the second morning after that conversation. We said, in the interim, many goodbyes, and made many plans to communicate and perhaps to meet again. Along with a new language, he was to get a new name, an Anglo-Saxon name. He acquiesced to my suggestion. Vito Abraham Fishbeingnola was to become Victor Fishbourne. Doctor Victor Fishbourne. And, although we agreed to send missives back and forth so I could be told of his progress through life, I thought we'd ever see each other again.

Chapter Fifteen

Around 1625, give or take a few years, a wandering priest came in and out of the poorest districts in Padua, Venice, and Florence. He had obviously taken a vow of poverty, since his clothes were frayed, his face was gaunt, and his demeanor bowed. He was generally introduced by parish monsignors, who seemed to treat him with some deference, as a man who would hear the confessions of the most depraved. He was called simply Father Cesco, and no other name, description, or validation was offered. He was often seen counseling those most in need of spiritual uplift; drunkards, cutthroats, robbers, those who almost never came to church. He would find them in their taverns and seedy rooms, in their whorehouses and shacks, and convince them to come to him for absolution. The parish priests were amazed that he had such rapport with such a group of rascals and mortal sinners. If he were a rich priest from Rome they would have thought that he was bribing the sinners to confess, but they had been told by the offices of the bishops themselves that he was just a poor, wandering priest, seeking his own redemption by providing it for others.

Excerpts from the memoirs of Salvatore Omar Fishbeingnola, c. 1626.

When I returned to Padua after my misadventures with my son Vito on the road to Commune de Tirol, I was already suffering with remorse. I knew I had done the right thing in sending him to England, but I also knew, somehow, that I would never see him again. I would shake this remorse by sending for an old ally, my informant Amadeo. He, if anyone, could tell me if my assumptions were correct; the ghosts of the Bellarminos were extending their curse upon me through that sycophant Montepulciano. This invidious priest's visit earlier in the year had been threatening. Was he somehow involved in making threats real?

Amadeo had aged, much more than I would have guessed, in the years since I had seen him last. Life in the dark quarters of Padua and Venice was hard. At least he was still alive. I bade him sit and poured some grappa for him. "Amadeo, my friend. How have you fared? I haven't heard from you in such a long time."

His was a faint smile. "I thought, after the death of Maffeo Bellarmino, you might not need my duties any longer. Certainly with the passing of Roberto you and your friend would be clear of enemies. I waited and watched, though. If you hadn't called me to come here, I was going to approach you anyway. It seems there are still those who seek you harm. And your friend, too. Since my life has become that of subsistence, I thought perhaps I could do us both a service. You for your security, me for my old age. I don't want to die a beggar, or on the front steps of a whorehouse. Yes, I was getting ready to seek your employ when you sent a message that you wished to see me. Fate draws lines that converge in the proper manner. I believe Da Vinci showed that in his paintings."

I thought of the crash on the way to the Alps, and of our miraculous escape, and of how the attempt on our lives helped my son and I make that difficult decision. "Yes, Amadeo, Fate guides us with some unseen hand, sometimes to a good place, sometimes to a bad. We understand so little about time, about why things happen when they do. For me, then, Fate is coincidence. Something seems important when it happens to you, and you wonder why it happened just then. In the end, all we really know is that it happened.: That I needed your help just when you decided that you needed mine is Fate. Or coincidence. My friend Galileo would say all it demonstrates is that we need better clocks." We both sipped at the grappa. "But what have you to tell me? And would you let me hire you once again, to keep telling me things I should know?"

"That's why I've come. And what I have to tell you, you need to know." He settled back in the chair, enjoying a plush seat. "Do you know of a priest, a Jesuit, by the name of Francesco Montepulciano."

"The very same person who came to visit me earlier this year?"

"Yes."

"The man who made veiled threats against Galileo, me, and my family?"

"That man."

"And why do you bring his name up now?"

"Because, Signore, he has, for several years, made regular visits to Tuscany, disguised as a poor traveling priest. He claims to hear confessions of only the most dissolute, the nearly irredeemable, and gives such people spiritual counseling. And something more. What he is really seeking is a band of thieves and murderers, paid to commit crimes of terror against a chosen group of people. That chosen group includes you and your friend Galileo. I learned too late of the treacherous coachman you hired, who acted under the hand of this priest. I would have warned you. I can tell you now that certain burglaries of upper class people, correspondents of yours, or you son's, or your mother, or Galileo, have not been for the trinkets that were stolen, but in search of incriminating documents, documents that would certify your background as non-Christian, documents that would certify that Galileo is in league with the devil, that he gets his ideas from demons who visit him in the night. Montepulciano has been unsuccessful in such acquisitions, so he decided upon an 'accident' for you and your son. I'm not sure if the word of your survival has reached him. When it does, the plots begin anew."

I wasn't surprised by this news. Rrather, I was glad for the confirmation. Something puzzled me, though. "You say he indulges in bribes when he finds a man willing and able to carry out these assignments. Where could he get the money? Surely the church wouldn't finance such evil."

"I wouldn't be surprised if old Roberto Bellarmino gave him some valuables before he died."

"Are you saying that Roberto was complicit in such things, from beyond his grave?"

"I doubt that. He probably told Montepulciano to use the wealth to carry on the work. Montepulciano would take that to mean by any means necessary. Montepulciano was not unaware of Maffeo's insanity. In fact, he copies it."

I took a strong draft of grappa. Amadeo followed my lead, so I refilled our cups. I ran my fingers through my thinning hair and asked, "And now you know of further plots against those I hold dear?"

"I know nothing firm, yet. If you want me in your service, I will find these things out. But, I have something more to offer. Turning the tables, if you will."

"Tell me, please."

"Do you remember a man called Guido? He was sometimes called Il Venezorino."

"I have known many Guidos. I know of no Il Venezorino."

"This Guido tried once, in the Alps, to kill you."

"I never knew that man's name, until now."

"And now you remember that it was he who sent you to me, to help keep you and your friend safe from the hatred of Maffeo Bellarmino."

"Yes."

"And that Guido disappeared for a long while right after he failed to do you in?"

"Yes."

"He went to join the German Catholics in their war against the Protestants."

"I know."

"You knew?" Amadeo seemed truly surprised.

"Yes. He vowed me to secrecy after I saved his life. He told me of his plans at the same time he was recommending that I hire you. I keep my vows."

"I appreciate that. It's one of the reasons I like working for you, and one of the reasons you can count on me."

"But, you were going to tell me about this Guido."

"Yes. Guido. Well, he was safe for a while. A few skirmishes way up north, nothing a man quick with a weapon couldn't get through with no more than a few cuts and bruises. Then he was involved in a major battle. He probably should have died. Almost bled to death. As it was, he lost the use of a leg. He still has it, but it doesn't work. So, neither does he. He's even more needy than I. And, he remembers you fondly. He would be a good resource for what I have in mind, since he's still capable of violence as a persuasive technique."

"You should know, Amadeo, that I deplore the use of violence as a means of solving problems."

"Even when the problems are someone who doesn't deplore the use of violence to solve problems?"

"You're not suggesting......"

"No. Not murder of any sort. Just persuasion to induce certain craftsmen to produce certain documents, certain artifacts. People to make certain testimony. Sometimes it will only take money. But money plus the threat of force are often bargains most people can't refuse."

"I don't see where you are going with this."

"Just this. Even Montepulciano would rather put together a civil and religious case against you and your friend, just in case terror doesn't work. I believe it is possible for you to do the same thing, as concerns not just Montepulciano, but Bellarmino as well. Never to be used, unless necessary."

"The idea of legal debate arises all the time. I have never considered it for actions that haven't even taken place."

"And that's what I've come to offer, Signore. A plan."

"I will listen, but not to a plan that includes violence, even to people who try to do violence to me. I have had some guilt about actions I've commissioned in the past."

"For you sir I've only met violence with violence. You have never commissioned a preemptory act."

I had to think on that. "I suppose that's true." When I realized that I had a little of the feeling a boy gets after a visit to the confessional. "So, what exactly is your plan?"

"First, I know many fellows who have been lured into the confessional by our Father Cesco. Some of them are the most hardened of cutthroats. Some have talked to me. They are witnesses to the real reason the beggar priest has made the rounds. He has bribed and threatened these men to perform illegal acts, burglary or assault, extortion and kidnap, to gather information on his enemies. Documents, activities proscribed by the church, secret Copernicans, secret Jews. He only wants the information. He lets them keep any swag they can find or ransom."

"So. Montepulciano is a bad man. I know that. Will testimony by a bunch of blackguards demonstrate that to the Cardinals?"

"Probably not. But it will get Montepulciano's attention. There's more. Of a different nature."

"I don't understand."

"Montepulciano is the major disciple of Roberto Bellarmino. He has devoted his life to seeing Bellarmino canonized. It is he who has gathered the majority of testimony on saintly acts and the history of miracles attributed to Bellarmino. He entered this information into the litany under the names of many others who wanted to be part of the construct but didn't want to do the necessary work.

"But among the witnesses to those acts and the experts investigating the miracles are doubters – even liars. You see, not all the people who were threatened or bullied or extorted were actually guilty of some secret offense. Some were upstanding, virtuous, and credible. They just fell prey to an unscrupulous man. Oh, perhaps they were already leaning toward supporting sainthood for Bellarmino and saw nothing wrong with exaggerating a little. Not everyone is as honorable as you."

"Do you think some of these people will renege?"

"If they know the man behind the threats could be neutralized by their complete honesty, I believe they would. You see, they needn't admit to mendacity. Just to making their memories more precise. I think a few such statements, properly documented, along with the songs of the cutthroats, could put Montepulciano on the defensive. That will enable you to bargain with him."

Amadeo had amazed me once again. The plan was intelligent and articulated well. Why wasn't a man who could scheme like that a Doge? I don't know the answer to that. Just let it be said that I trusted Amadeo to be working for me, and not another implant of the priest Montepulciano, as the coachman had obviously been. Moreover, Amadeo's mention of this Guido or Il Venezorino or whatever he called himself, convinced me of the validity of his intentions. Perhaps Guido still felt indebted to me.

Perhaps he needed some sort of livelihood. In any case, he was a real person, a man of deeds, who would probably have just one master at a time. How sure could I be of these men? Amadeo saw my hesitation. "Yes, Amadeo. I'm thinking about this. I feel confident in your loyalty. Should I? It would be very clever of Montepulciano to send an agent offering to work for me. I don't think you would do such thing."

"Oh, make no mistake. I have had feelers from that priest, just because of my lifestyle and the friends I have. I have never gone into one of his confessionals, however. I am going to Hell, Signore, that's true. But I must do it in my own fashion. There is honor among thieves, just as there is dishonor among you legitimate folk. You were always honorable with me, and that you kept Guido's secret after he tried to kill you says a lot. It is why you can trust him, too. And we will be all you need." With that, Amadeo proceeded to the details. I thought it a good plan, but I hoped I would never have to implement it.

***

Before he left, Amadeo had one more question. "No one has heard anything of your son after the 'accident', nor has anyone seen him. Was he hurt? Is he still up there? Is he even alive?"

I gave Amadeo my most steady look. "As far as anyone is concerned, my son died in the carriage crash. As far as Rome is concerned, as far as Italy is concerned, my son died in the carriage crash. That's all I will say."

"As you wish, Signore. That will be the word on the street."

In making my pact with Amadeo, I didn't lie, and I didn't solicit murder. I did, however, have to sink to Montepulciano's level to fight him. That's always the way in human conflict.

Chapter Sixteen

In 1632, Galileo published *Dialogo sopra i due massimi sistemi del mundo, Tolemaico e Copernico.* (Dialogue concerning the two chief world systems, Ptolemaic and Copernican.) For purposes of this history, the most important thing to note is that it was published in Italian, and not Latin. Apropos that, there are words in Italian that are congruent to words in Latin, but may have very different meanings idiomatically. Another item of note: The first principle of Einstein's Special Relativity Theory is that no experiment in a closed laboratory can yield information about the motion of that laboratory. In the dialogue, Galileo points out that a physics experiment in a closed ship's cabin cannot be used to tell if the ship is moving or not. It might have been a little too soon to try to spring relativity onto the world.

(Author's footnote, not at the foot of the page, and without citation, other than 'trust me.' By the way, an old Sicilian aphorism says 'Never trust someone who says 'trust me.' This is not quite as paradoxical as saying 'Always trust someone who says 'Don't trust me.'
Anyway, the footnote(s):
In the pages below, Galileo will drop a couple of names. Here is the background for them.
Christoph Scheiner was a Jesuit theologian and astronomer who claimed description of sunspots prior to Galileo. There was a dispute about who

mentioned them first in published works, and each proclaimed the other a plagiarist. It is probable that neither was first to notice them, but both observed them well and described them more closely to modern thought. Galileo's data about their motions over time may have been better. They showed that the spots were not transits of Mercury, which was a blow to Copernicus' argument describing the Sun as the center of the universe because it was so perfect.

Orazio Grassi was also a Jesuit astronomer. He had a long running dispute with Galileo over the nature of comets. Neither understood them very well, but Grassi was a devotee of the Tychonian Universe, devised by Tycho Brahe, the Danish astronomer. This model kept the Geocentric Universe, with the Sun and stars orbiting Earth, but with the other planets orbiting the sun. Interestingly enough, Tycho's math was better than that of Copenicus in this very complicated model. End footnote.)

Excerpts from the memoirs of Salvatore Omar Fishbeingnola, c. 1632

It was late at night without notice that Galileo came to visit us. The servants had gone to bed and Anna and I were chatting over small glasses of wine. His knock was so careful, so light, that we wouldn't have noticed if we weren't in a cozy study at the front of the house. It was our favorite place for our private time together, and we often sat close enough to touch if that was a proper of expression for a given idea. It was Anna who said, "Is that someone at the door?"

"You stay here. Let me check." There was a small port in the corner of the room that could give a view of the front stoop without the viewer being seen. I peered through. It was Galileo. When I opened the door to him I started to ask the obvious question, but he put his finger to his lips and hurried past. He headed for the little study. Anna and I shared our private place with him on occasions, so he knew by my rapid appearance at the door that it was the study from which I came. There were only three chairs in the place. He took one. "Can I get you anything? Something warm? The servants leave a pot on the fire." He looked as if he were shivering.

He shook his head. "I need nothing for the body. It is my soul that is icy. It feels a chill from my enemies, and I don't know why they have waited this long just to rehash old arguments. Salvatore, what have I done recently that is so different from my actions, my thoughts, my philosophy through all these years?" He answered his own question. "Nothing."

Wordlessly, Anna left the room, and returned with a blanket, which she put about his shoulders. "You must remain calm, Galileo," she said. "You must keep your head about you. You are smarter than they, better able to present your thoughts, always bearing better thoughts. Don't waver now."

I understood Anna's support for my friend,

but I disagreed with her on the proper approach. She had expressed her admiration for him many times, her conviction of his genius, and had always encouraged him to stand by his beliefs. I, on the other hand, had always advised discretion. It was one thing to invent gadgets that measured time and temperature and pressure better than anyone, and it was fine to tie together, in an intricate fashion, mathematics and physical science, but it was quite another to flaunt the results in contradictions of church teachings to powerful church factions. I had always encouraged Galileo's science. I had always been less than encouraging about his publications. And I had always discouraged, as gently as I could, his disputes and antagonisms with his church. I had feared it would come to a head, as it had in 1616. Galileo's presence in my study suggested it had done just that.

Anna and I waited patiently. When he was ready, Galileo said, "Some of this I could understand, if I could believe the gossip. I've had my troubles with Scheiner over the sunspot thing. I plagiarized. He plagiarized. I don't care. I would willingly grant that he observed them independently. Who was first is lost in the years. What they show is the real debate. And Grassi. I know I put him down in the Assayer. No more than he has put me down in his works. That I am more widely read and believed is just the way it is. Can these men be so vindictive?" He shook his head. We waited. It was obvious he wanted a response.

"They have the Pope's ear," I said. "And this

business of Simplicio. It is obvious to everyone that Simplicio represents the Pope in the dialogues. It was a bad choice of names for the loser in the Copernican debates."

"Simplicius was a great thinker. His name indicated a straightforward person. He's Simplicio in Italian."

"It also means 'simpleton' in street Italian. How many times? How many times have I told you? Write in Latin. Don't write for the streets. The people don't care about the Earth moving. Many of them think it's flat, and you can't convince them otherwise. They do take great delight in someone making the Pope look like a simpleton, and calling him one in the bargain. It will be very difficult to get his other ear now."

"Barberini is my friend. He'll listen to me."

"He is no longer Barberini. He is Urban VIII. And you've hurt him."

Galileo slumped in his chair. "What has happened? Even since Kepler died I've been losing allies. I *did* report the sunspots first, but I was willing to share with Scheiner. I did concede that Kepler might be right about the tides. He was a friend. Why not Scheiner? And Grassi. He's the one who should be insulted by the lambasting in *The Assayer,* but we put his ideas about comets aside over ten years ago. Does he still hold a grudge? The Laws of Nature are about mathematics. I've shown that time and again. That these people continue to argue shows that they don't understand. That they vilify me with false conjectures proves that they are not good men, let alone qualified to be called scientists. But I have to tell you, my friend, for once I am at a loss for a course to take. Except to hold firm to my principles."

I sat straight in my chair. "Old friend, I would never suggest you let go of your principles. They are you. What I do suggest is that, for the purposes of your survival, you soften your stance. At least publicly. Then you can continue your work. And you are one of the few people in the world whose work matters to all the rest of us. You are one of the few people whose work can change history. Your burden is that you may not be around to enjoy that satisfaction. There's little that can be done about that. Now I...."

Galileo jumped from his seat, frowning. He didn't want to hear this.

I remained calm. "No. Hear me out. I have been thinking about this for a long time. Preparing for it."

He glared at me. "Again you go about protecting me behind my back?"

"My friend, it not entirely altruistic. There is some self-interest involved. Your enemies are mine, whether I choose them or not. Whether you like it or not. I am also thinking about myself. I made a pact, you might say with the Devil, some time ago, just for this contingency. I will put it into play, but you must promise me that when Urban calls on you to recant, you will. You don't have to mean it. I will ensure that everyone knows you didn't, but the Inquisitors will have the satisfaction of your recantation, so they won't torture you or burn you at the stake. Your fame should preclude that anyway, but I want to increase your chances of escape from harm. Will you trust me?"

Galileo sat back down. "You swear to make it known that I truly have shown that the Earth revolves about the Sun?"

"I swear." We didn't talk much after that. Anna bade Galileo stay the night, but he was too preoccupied to accept. We understood.

***

I didn't like traveling to Rome. The countryside was pretty enough, but the roads were rough and one spent a good deal of time on them. They were probably better when Rome belonged to the Romans; now, after the Dark Ages, they had fallen into ruin, in some spots becoming porous mud holes, in other places the site of some exotic boulders. True, the roads in the Tyrol weren't any better, but the Alps were the Alps. I couldn't fret about the road, however, because of the importance of this trip. An agent had tracked down the whereabouts of Father Francesco Montepulciano, and I needed to talk to the man. It was he who would be my vehicle toward safety for Galileo.

The scoundrel! He was now posing as a parish priest in the most teeming quarter of Rome, even though he was truly an agent of the Index. It was his job to spy on businessmen who might be too successful and not generous enough in their Sunday offerings. Tidbits about wavering faith could bring generous tithes to the Church. It was just one of his jobs. His main occupation was providing information on suspect individuals to the Inquisitors. He carried out his duties assiduously, and was trusted by many who had the ears of Cardinals and even the Pope. My agent told me where I could find him at ten o'clock every weekday morning - having a bit of pastry at a small shop on the Via Pomposa. My agent was correct.

Montepulciano's eyes widened as he saw me approach. Before he could speak, I sat at his table. "Father," I said, without any amenities, "I won't take much of your time. You should listen very carefully, and look closely at what I am about to show you." I removed the pouch strap from my shoulder, laid the pouch on the table, and opened it. I removed several documents and spread them before him, in two groups. "This group of papers contains sworn confessions of various thugs whom you hired to perform various deeds, each of a heinous nature. Any one of these could be dismissed as slander. Together, they describe a priest who is a less than a holy man."

Montepulciano drew back his shoulders. "How dare you accost me, without even a civil greeting, and threaten me with nonsense?" His eyelids squeezed down to slits. "I've always had my suspicions about you. Now they're confirmed. You'll soon learn who is condemned as less than holy. Do you think, before any council, you and those papers, if they're even real, will be believed over my word? I'll denounce you as a Jew and a heretic. I'll see your whole family ruined and put to the rack." I didn't have to look at him to feel his rage.

I remained calm. "Yes, there might be some debate, and yes, you would probably win, given the prejudices that would be in your favor. Still, your patrons would forever have their doubts about you. Whatever authority props you up now would abandon you, simply because of that doubt."

"Never."

"And that's why I have this second group of papers. And these are from holy, pious people, most literate enough to sign their names. What do you suppose these papers might be about?"

"I have no idea."

"They pertain to certain miracles that are associated with your demigod, Bellarmino. I know you wish him to be a saint one day. That will take verification of miracles, and proof of a saintly life. These papers demonstrate that either the miracles didn't occur, or the testimony you have gathered was false. They also indicate certain less that saintly traits exhibited by your demigod, things that would automatically exclude him from consideration for sainthood. And, as I said, these are from pious people, upstanding people. Even several priests. One in fact from the Society of Jesus, just like yourself. I can guarantee that I can sully Bellarmino before consideration ever begins, and you will spend the rest of your life trying to prove these documents are invalid. And when you die, you still will not have accomplished that. And you will never be able to connect any of this to me."

Montepulciano was shaking his head, mindlessly putting a piece of pastry in his mouth. He tried to speak, spitting crumbs at me. He calmed. "Why would you do this? What could a Jew have against a deserving man becoming a saint?"

"In the first place, I'm not a Jew." This was arguable, but not at this time. "It's not about my religion, it's about yours, and I really don't care who Catholics make a saint. What I do care about is who you burn at the stake. And it won't be Galileo. And he won't go to the rack, or be thrown into a dungeon."

"So, you do this for yet another heretic. But I can do nothing about that. He has made his own problems, and he must answer for them."

"Don't be so modest. Of course you can do something. You have the ears of Scheiner and Grassi, and they have the ear of the Pope, and of the Inquisition. And here's what you can whisper in those ears. Galileo will recant his Copernican views. In return, he will be given some light punishment. House arrest perhaps. He's getting older, doesn't need to move about too much anyway, and he can carry on experiments wherever he might be. This will show the world that the Pope appreciates the value of my friend, and the Church will have the satisfaction of his recantation. The general public will think the Church is charitable and forgiving, and also right about how the universe works. My friend won't suffer needlessly."

"You give me too much credit if you think I can play such a part."

"If you can't, then Bellarmino isn't really a saint. If you believe he is, you must pray that he guides you in this task. And if he isn't, I will see to it that the world knows."

"I curse you."

"Curses I can live with. It's assassins and slanderers who give me trouble. And that brings up the other condition. Me and my family are to be left alone. That is a simple request, and something that is entirely up to you, since, unless you have already put

my name on some list, I am unknown to the
Inquisitors anyway. Do you understand? Leave me
and mine alone."

Montepulciano, still without noticing what he
was doing, put the last piece of pastry in his mouth.
He nodded and sighed. "I will pray to the holy
Bellarmino, who even now is in heaven, observing us.
You especially. If he sends me a sign, and gives me
the power, your friend Galileo will be spared."

"And me?"

He hesitated. That pause told me he was
about to lie. "And you."

Chapter Seventeen

In 1633 Galileo was called before an Inquisition. Why? And why then? He suspected that his *Dialogue on Two World Systems* had hurt the feelings of Pope Urban VIII, who up to that point had been friendly to him. He also suspected that two powerful opponents, one of whom he had defeated in debate and the other of whom felt that he, Galileo, had plagiarized him, had convinced the Pope that Galileo had made him a fool in public. This was never shown to be true. It was also never shown to be false. For some reason, also not known, the two men in question, both priests, had, it was rumored, changed their attitude toward Galileo, subsequently causing the Pope to think more kindly of him. In addition, the Inquisition discovered that much of the evidence about non-public blasphemies alleged to have been perpetrated by the scientist had disappeared, and the chief witness against him had declined to testify.

It was customary in an Inquisition, especially one in which the defendant was known to be guilty, to follow a fixed procedure. The evidence was presented, the verdict given, and the defendant was then given a description of the instruments of cleansing, i.e. torture. The description was wasted on Galileo; he could have invented more hideous machines, so it was easy for him to imagine the ones already in existence. The person in question was then given time to think about it. If the heretic did not confess or recant, the next step was to take the person to the chamber of horrors itself, and demonstrate a few cranks and gears and sharp things. There then followed a period of consideration of what had been seen. If the heretic still did not yield, the instruments were actually applied. With Galileo, the procedure did not even reach the description stage. To everyone's surprise, after the charges were read and his published works presented, he recanted. Before the world he disclaimed belief in the Copernican System. No one was sure why he was suddenly so pliable, but Urban was satisfied, Scheiner and Grassi were mollified, and only a little known Jesuit whom everyone called Father Cesco seemed agitated. The Inquisitors agreed on house arrest for an indefinite period, at a place of Galileo's choosing.

A rumor began to float, mainly in the underworld, that Galileo muttered, as he left the chamber, "Still it moves." This, of course, referred to the Earth, and meant that his recantation was a lie of convenience. That he actually said this has never been proved. In fact, this information didn't reach proper historians until many years later.

Excerpts from the memoirs of Salvatore Omar Fishbeingnola.

I was justifiably proud of myself. I had sold my friend on the idea of a false but public recantation, and I had convinced the monster Montepulciano to call upon his powers to bring about the most lenient punishment for that friend. Whether he did it by prayer to a saint who only existed supernaturally, or by more mundane exhortations to men who were mortal but who could actually wield influence, was of no concern to me. It got done.

Because Galileo knew how it got done, it was easy to convince him to spend some time with me, at least until he could find a place of his own, one which would seem like a home and not a prison. I made one of our spare rooms into his bedroom, another into a study for him, and yet another emptied to receive his laboratory equipment. He could carry on his work unimpeded, and I could see to his safety. I didn't trust Montepulciano, and the streets were full of the young and unemployed, living on the edge of poverty, and willing to commit any act for a few gold coins.

I convinced my friend to stay with me for at least a year. In that time, he realized he would prefer to retire at his country house in Arceti. I wasn't only thinking of him when I asked him to be my guest. Anna loved having him, loved having someone else to take care of. For my part, Galileo helped fill a void in my heart, a void left by the departure of my son, whom I now knew only by sparse correspondence smuggled back and forth by friends. That my son had prospered in England was small compensation. He was physically gone, for what I considered forever. At least I had Anna. Galileo had a daughter in a convent. Although she was in Italy, she seemed farther away than my son in England.

And there was the other 'son' – Federico
Fangio. Fang. I hadn't heard much about him in the
years since he had been shunned by his family. My
informant Amadeo, my co-conspirator in the
Montepulciano business, had kept his word to Vito.
Although Amadeo had worked for both of us when
Vito was in Padua, helping protect Galileo for me and
helping protect Fang for Vito, Amadeo had
maintained a confessional like sanctity. He had, as
sworn, told neither of us about the other's affairs.
Even now, with Vito far away and obviously not
dealing with Federico any longer, Amadeo was silent
on what he had done or might be doing for Vito's
friend. I didn't pry. I just took notice of stray bits of
information as they came to me randomly.

As I understood it, Federico had become such
an accomplished craftsman that he could truly be
called an artist. He was employed, secretly, by
people who dealt in ornate glass, either for windows
or for three dimensional objects of art. I was told that
several of the most respected sculptors in the region
employed Federico to complete or even completely
create works for which for which patrons of the arts
paid handsomely. Some people said he had a
beautiful wife and two children who were educated in
Jewish *schules.* There were no harrowing tales of
persecution as far as I knew. This let me turn my
attention to finally getting my friend Galileo securely
settled, once and for all.

Chapter Eighteen

In 1634 a delegation headed by Sir Thomas Heedon arrived by ship at the port of Venice. Sir Thomas, Eighth Earl of Liverswil, had become fascinated by the notions of the laws of the world and of the universe contained in the studies of Galileo. He had some vague idea of forming, in England, a royal society for the improvement of natural knowledge. His idea was gaining no traction in his own country, so he felt he needed to talk to Galileo himself, to pluck from his mind the inspirational phrases and didactic tools that would convince people that science was the path to truth. With him was a young man who had made a mark in British society by displaying an uncanny knowledge of Galileo's experiments and the intentions of his philosophy. The young man had appeared in London in 1628, leaving a thriving medical practice in Lincolnshire to attend to business activities in London, and to begin teaching and experimenting at Cambridge in what was now starting to be called Science. The name of the young doctor/natural philosopher was Victor Fishbourne. Sir Thomas was glad to have the young man along. Not only was he brilliant, he was amazingly fluent in Italian. Sir Thomas himself had only recently learned the language and still struggled a bit with it, although he was adroit in scholarly Latin.

As they were preparing to leave the ship, Victor addressed the Earl. "Sir Thomas, I have a small favor to ask. Rather, a plan, if you will." This was delivered in his rather pleasant upper class but not foppish accent.

"Of course, Victor. I'm always willing to abide your ideas."

"It has occurred to me that if we are to gain the most from our discussions with Maestro Galileo, we might best do it separately. You and I have slightly different approaches to this task. You are motivated by the founding of a society of thinkers and scientists, I by wishing to do further experimentation and teaching in the natural sciences. We have different purposes and would therefore have different questions and different perspectives. We might interfere with each other, break each other's train of thought, misunderstand the subtly. One might want greater insight into a particular mechanism, a mechanism of less interest to the other. And, in the interest of honesty, there is certain business I would like to attend to in Padua."

"How long will this business take?"

"That depends on how well my agents here in Venice have administered my interests. I would say at least two days."

"I knew you had extensive financial interests outside your medical practice, but I wasn't aware that you had business interests in Italy. I suppose I should have guessed, given your fluency in the language. Still, I was counting on that to be of help when we met Signore Galileo. Some of his best work has been written in Italian, and I've only read translations. I need some help with the nuances."

"As I understand it, Sir Thomas, Maestro Galileo's current host, Signore Fishbeingnola, knows the work quite well and does well enough in English to be of great assistance to you. I'm told Fishbeingnola has been a friend and confidant of the Maestro for many years. Since childhood, in fact. I think he is better prepared than I to divine the Maestro's true thoughts."

"That's well and good, and I'll go along with your plan if you feel your other business is urgent. But I'm not going to stay in Padua for more than two days. I've heard of a place in the Tyrolean Alps, Commune de Tirol, that is supposed to be a marvelous resort. I wish to relax there for a bit before we return to England. Ever heard of the place?

Victor thought for a moment before answering. "No. And my Italian agents haven't mentioned it, although they are well traveled, at least in the Venice Republic."

"I'm sure you'd like it there. I'd like for you, when you're finished with your business and with Signore Galileo, to join me there, rather than meet me back here in Venice. Would you mind the trip? Could your agents direct you to the place?"

"I'd be very happy to meet you there. I'm sure my agents can put me on a carriage bound for this Commune de Tirol."

"Good. Good. Just don't miss your interview with Signore Galileo. I understand he is leaving the house of this Fishbeingnola fellow to serve his sentence at his country house. He might not be so receptive to visitors in the future. And who knows about the health of anyone seventy years old?"

"I will make it to the Fishbeingnola house. I promise."

"Good. And I will see you in Commune de Tirol."

When Sir Thomas said that, the words 'full circle' came to Victor Fishbourne's mind. In English.

Excerpts from the memoirs of Salvatore Omar Fishbeingnola.

We, and by that I mean Anna and I, would have been happy to have Galileo live out his life in my house in Padua, but he was adamant about moving to Arceti. I understood why that would be. The house was spacious, quiet, and away from prying eyes. He had permission to leave it for medical attention and for trips to acquire personal items. It would feel less like a prison than any other place he could imagine, and while he would be losing the tender care of my beautiful Swede, he could hire a relatively competent housekeeper and cook. I would make sure they were honest and loyal to him, and as usual, I would pay them a little extra to try to approach the care that Anna provided here. I say 'as usual' because I wouldn't tell my friend I was doing it. I thought of hiring a watchman, but I saw no way of keeping that a secret, and I felt that Montepulciano's time of harassment was past anyway. We were all getting too old for such childish games. Moreover, I was keeping my promise to Montepulciano. He could go about altering history to try to set up the beatification of Bellarmino and I wouldn't interfere. I didn't think he would live to see it, let alone canonization, but it gave him something to do. It's interesting that only men are priests because priests, especially Jesuit priests, are like women - they have to have something to do, or they get cranky. The difference between priests and women is that women, for the most part, will keep looking for something to do until they find something useful and usually benevolent. Priests will just pick the first thing that takes their interest, good or bad, because they are also just men. Perhaps all priests should be women instead of men.

Galileo was more than ready to leave, but he had granted an interview with a British noble who was trying to form some sort of society in England. The man wanted to probe my friend's mind, hoping to find that vein of imagination and expansiveness that would lure the British Upper Class and Intelligentsia into forming a club, an exclusive and intellectually snobbish club, which would advance this thing called

Science, which my friend seemed to have invented.

The interviewer, a Sir Thomas Heedon, arrived accompanied only by a valet. We had expected an interpreter, since there was a small language barrier, but I agreed to do my best to translate. When I inquired about the whereabouts of his translator, Sir Thomas' answer froze my heart.

"You mean Doctor Fishbourne? I'm sorry to say he had important business here in Padua and extends his apologies. He's really quite fluent in Italian and would have been useful. He promised me he would be here in two days to continue the interview with Signore Galileo."

I hoped I had kept the involuntary gasp from his notice. Apparently I had, for he immediately asked after Galileo. I realized that I was merely a facilitator of a conversation that didn't really involve me. I was glad of that, because knowing that my son was close by occupied nearly all my thoughts. I would stumble through any translation that might be required, but my head would bear the burden of anticipation. Vito hadn't come for some 'business', he had come to see Anna and me. And he wanted to do it in private. He knew that Galileo's departure was imminent, and he wished to be alone with us. And to avoid an awkward meeting with Galileo. You see, I had never told anyone what had happened on the road to Commune de Tirol. I wanted everyone in Italy to assume he was dead. The carriage accident wasn't planned. The disappearance of my son was, and the accident, a failed assassination attempt, actually gave me the perfect story. Only Anna knew the real outcome of that event. Galileo didn't need to be troubled with it.

Under any other circumstances, the two days of interviews would have been fairly interesting. Sir Thomas was a reasonably intelligent fellow, and certainly a partisan of science. There was hardly any need for interruption for translation, since they seemed to get along quite well in scholarly, if poorly pronounced Latin. I should have followed more closely, and would have, if my mind hadn't been filled with thoughts of this surprise visit by my only offspring, an offspring I had given up any hope of ever seeing again. Had he changed so much as to be unrecognizable to every old friend or acquaintance who might cross his path?

At the end, Sir Thomas Heedon seemed satisfied, although I was at a loss to see why. I presume he had some greater understanding of astronomy and of physical science, perhaps of the ascendancy of mathematics in these disciplines, but to what purpose? This talk of forming a society of some sort for the furtherance of these studies seemed naïve. It would be a long while before English society would accept this as anything more than an indulgence for the upper class. After all, there were no secular universities to be found. In one of his letters Vito (I still called him that) had thanked me for his Catholic upbringing. It had helped him insinuate himself at Cambridge, an institution which, although Anglican, held all the Trinitarian beliefs of Catholicism. He fit right in, so when he began teaching science and so many bright young men listened to him, no one felt heretical.

I bade Sir Thomas goodbye, wishing him a safe continuing journey. I almost gasped again when he told me of his plans to visit Commune de Tirol, and even meet his fellow traveler Victor Fishbourne there. After, of course, Victor had his own interview with Galileo. The ironies were becoming almost more than I could bear, but I was now certain that Vito would appear the next day.

That was not to be the case. It was two days after Sir Thomas left, and one day after my friend Galileo left for his villa at Arceti, that my door was opened, not to Vito, but to my old informant, Amadeo. He was excited, almost breathless, and while he bowed politely to Anna, it was obvious he was impatient to speak to me, alone. I took him to the little study in front of the house, and after I poured him some grappa, he told me the most amazing story. Even he did not know how amazing it was.

Chapter Nineteen

Amadeo's Story as told to Salvatore Fishbeingnola.

As you know, Signore, I had almost forsaken my duties for you. You yourself told me that Montepulciano probably no longer represented a threat, and with your friend Galileo ready to retire to his country place, your family all dead save for your housemate Anna Bolen, you felt you could live your days in peace. Still, I am in many ways a more suspicious man than you. I have known many evil men, and I have never known the truly evil to change. Except for things done in your employ, I am still the same thief and cutthroat I have always been. There was no reason to believe that Montepulciano would become honorable just because he kept a bargain with you on one occasion. No. He is an unforgiving man. And a bad one.

Because I knew this, I always brought up the wandering confessor in my conversations with my friends in the darker side of the city. By now, almost all knew that he was more than a simple priest living a vow of humility and absolving the sins of the misbegotten. He had hired too many ruffians, even assassins. He had planned and paid handsomely for burglaries. In those, he had let the criminals keep the valuables; he only wanted certain documents. He had used those things to extort people into testimony he desired, or into the performance of other foul deeds. Granted, he was less a topic of conversation than in former days, and he had been totally inactive for some time. Then, he returned to the alleys of Venice, of Padua, of Florence. He was up to no good, but his activities were known only by those of us who were also up to no good. No one of us would bear witness against him, save for a few, like me, being paid to inform, then only to private parties for private reasons. You look surprised that other rich men keep informants in the underworld. Do you think you are the only man of means who is willing to pay a small price for security?

By now this ill-begotten priest's eyes had taken a certain vacant look upon which everyone remarked. They seemed to have no soul behind them, only a mind in chaos. Some people, including some of the most depraved, were fearful of him. They saw the Devil in his shadow, and his absolutions were considered condemning rather than redeeming. Only the most desperate would even listen to him, and most of those walked away from schemes so heinous as to make a hardened criminal shudder. It was conceded they were the ravings of a madman, and would have in the end been ignored if they hadn't contained a consistent thread –one tied to you and to Signore Galileo. So, in the underworld it was known that Father Cesco was going to try to harm you and your great friend. Informants saw no need to tell their employers things that didn't concern them. The usual miscreants saw no reason to tell anyone; for them, it was a form of entertainment, a real life drama unfolding before their eyes. As for me? If you will recall, I have come to you several times in the last year and warned you that you might be in danger. You were patient with me, but you never became alarmed, because I could present you with no specifics. I didn't mention the name of the villain because I wasn't sure if he was capable of mischief, and you seemed at peace. I vowed to stay on my guard, though.

Then, two nights ago, it came to a head. Father Cesco, whom you know as Montepulciano, had finally succeeded in unfolding an acceptable plot for two hired assassins, one that would pay them handsomely and put them at little risk of discovery. It all centered around you helping Galileo move to his villa at Arceti. It was to work much like the accident that killed your son and almost killed you, years ago, on your way to Commune de Tirol. Only this time, the coach hired by you and carrying Galileo to Arceti would be driven into a ditch and overturned. You may or may not know that just before one reaches Arceti there is a sharp turn in the road, to avoid a small, dry canal. There is hardly ever anyone on the road there, especially after dark. One of the conspirators would be the coachman, jumping from the coach and letting it run into the ditch. Montepulciano, it seems, reached into the past for a plan already used once. Now, however, he wanted insurance. A second man, waiting in the ditch, would jump into the overturned coach and ensure that Signore Galileo, if not dead, was at least unconscious. The coach would be set afire, as if by its own lanterns. Montepulciano could consider Galileo burned at the stake, as a tribute to his idol, Bellarmino. He would then turn his attention to you. No one knows what plot he was hatching toward you. At least, I don't.

Obviously the plot didn't work as expected, for you know that your friend arrived safely in Arceti. I can tell you how it was foiled, but I can't give you the identity of the man whose counterplot foiled it. I only know his aide in that counterplot. Do you remember a man named Guido, a man whose life you saved, a man whose whereabouts you kept within your breast for years, a man who helped me help you prevent Montepulciano from harming Galileo. Guido came to me. He knew of the plot, a thing that I myself would have discovered, perhaps too late. He also knew of a counterplot, a coup that I never could have accomplished, at least not in time, and with such justice. Here's what he told me.

He had been contacted by a man whom he did not know, a man who came from the Jewish quarter but seemed of noble intentions. The man had a companion who stood off from the meeting. Guido described that man as resembling you, but much younger, young enough to be your son. That, of course, wasn't a consideration, since we all accepted that your son had died on the road to Commune de Tirol. This first man had his own contacts in the underworld, and had learned of the plot from a lowlife who had almost accepted the task, but had bargained for too high a price. This man, who must be a good friend of yours, and whose name we never learned, had conceived of a way to prevent Montepulciano's plan from coming to fruition, but he needed help. He had been steered to Guido, who had never lost his loyalty for you, through all these years. Guido came to me to enlist my aide, and since I remain in your employ, I agreed. I should say at this point, Guido was paid handsomely by the stranger. He probably would have accepted the task for nothing. And I expect no further reward than to stay in your employ.

The counterplot was simple. The man would hire another coach to take Galileo to Arceti. We would, however, allow the plot to proceed as scheduled. We got words to the assassins that Galileo would be traveling incognito, wearing a hood and letting protrude from it false black beard. They were to strike quickly and leave just as quickly after they had set the fire. These murderers were glad to concede haste. They wanted to be paid and to escape.

On the morning of the affair, an older woman offered to help serve pastries at the shop Father Cesco had been frequenting for such early repast. They accepted her offer of help, since they knew her from her tavern days in Venice. You know the woman. Bufana. The one with the ample breasts. She was easy to enlist, when the stranger made it known to her that Father Cesco was a worshipper of a Bellarmino. She put a drug in Montepulciano's catnip tea. Not much. Just enough to make him drowsy. After he left the pastry shop he wandered by an alley where

Guido and I awaited. He was weak and easy to
manage, and no one was about to come to his aid.
We took him to a small house the stranger had rented,
and there we kept him all day, until it was time for his
coach ride. We forced him to drink a much stronger
dose of the drug, and he became unconscious, a loss
of consciousness that would last for hours. He would
appear to be sleeping. We dressed him in a long
cape, with a hood, and placed a false beard around his
neck. All one could see at a glance was a hooded
man with a black beard. We took him to the coach in
a cart, covered in old rags. I engaged the coachman
in idle talk to distract him while Guido loaded
Montepulciano onto the coach. We informed the
driver that Signore Galileo was ready to depart, and
depart they did.

I must tell you, Signore, that the most amazing
thing is that it was so simple. The people at the
pastry shop were glad to see Bufana again, and never
suspected she had nefarious plans. When we
kidnapped the weakened Montepulciano he didn't
utter a single cry. I don't believe anyone near that
alley would have cared if he did. When we drove the
cart through the streets of Padua we hardly received a
second look. When we loaded our cargo onto the
coach, the coachman was all business about
departing, but not about checking his passenger. He
was just the paid driver and carriage toppler. And the
assassins believed that Galileo would be the
passenger in the carriage, and wanted only to set the
fire and collect their pay. I swear, Signore, there was
no suspense. The stranger knew exactly what he was

doing.

The outcome? The coach plunged into the ditch as expected. It was consumed in flames. There was a corpse in the smoldering wreck, but it was unidentifiable. The villains have yet to be paid, and are searching for Father Cesco, who seems to have gone missing. Some people, who were disposed, for one reason or another, to help Montepulciano in his quest to sanctify Roberto Bellarmino, have reneged. That quest has been set back dramatically. And the stranger who made all this happen? No one has seen him since he last spoke with Guido and me. He remains anonymous, and his motives remain unknown, but it seems certain he has a love for Galileo. Or for you.

Chapter Twenty

Excerpts from the memoirs of Salvatore Omar Fishbeingnola, 1633

Things had happened so suddenly. A visit by Sir Thomas Heedon, a hint that my son Vito was in Padua, his puzzling failure to arrive in time to interview Galileo, Galileo's departure for his permanent move to Arceti, and finally, Amadeo and his startling story of treachery and perhaps justice. One of Guido's mysterious strangers was undoubtedly Vito, but that raised more questions than it answered. The most pressing was: would he appear at my door soon? If he would only do that, he could answer all my other questions. How could he possibly have accomplished such a complicated plan without enormous preparation? Had he come from England on other occasions without telling me? What else had he accomplished about which I knew nothing? I needed to talk to him, post haste. Yet, I could do nothing but wait and hope he would come, and come soon. I could think of nothing else.

And my son Vito did come soon. The morning after Amadeo told me of the misadventures which almost befell Galileo, and which did befall Montepulciano, Vito came to my door. The servant who answered almost didn't admit him, since he was dressed in rather plain clothes, certainly not befitting a gentleman. He would have been turned away as an unwarranted solicitor if I hadn't been riddled with anticipation. I was at the door just after my servant answered, and knew Vito immediately. I dismissed my man with instructions to fetch Mistress Anna, and ushered Vito into the little study, where so much in my life took place. We were hugging joyfully when Anna joined us, and she greeted Vito in the same way. We settled in comfortable chairs in a circle, Anna and I with expectant expressions facing Vito. He knew what we wanted, what questions were hanging there as if supported by the dust in the air of the study.

My son, Doctor Victor Fishbourne, explained all.

"You know that I love you both very much, and I know this visit and all that accompanied it are a surprise. It had to be that way for many reasons. These lessons I learned from you, my father. One must, all his life, remain watchful for the safety of those he holds dear, and he must not let them know of his concern, of his precautions, so that they remain not only safe, but also unafraid. It's the duty of an honorable man to guard the true treasures of his life in the vault of his consciousness. Gold can be secured in a chest and hidden away. Loved ones need to be free and happy. It is that code, taught to me by you, that caused me to act as I did. Now I will be glad to answer any questions I have created."

"Surely," I said, "you have planned this for some time. But how could you know? And how could you arrange things so precisely? I will grant that you performed in much the same way I would have, with the exception of the actual murder of Montepulciano. I would have left him in the carriage with a few broken bones and second thoughts about his obsessions."

At this, Anna jumped into the conversation. "What's this about Montepulciano and murder? What have you been keeping from me?"

I had been waiting to tell her about the results of Amadeo's visit. I was hoping Vito would arrive and the whole thing would become clearer if the two of us explained the situation. That I should have prepared her better was clear, but it was moot. "I was going to tell you Amadeo's story this morning, dear Anna. You'll hear it now from two sources, along with certain questions for which I need answers myself. Don't be angry with me. I just didn't want to give you a skeleton when the whole body would soon be available."

"You're forgiven." Then she added, "Providing Vito and you can satisfy my curiosity."

"I hope we can." I turned to Vito. "First, was murder really necessary?"

With a grim smile he said, "I suppose you could call it murder. I call it killing a mortal enemy. Did he ever stop trying to do you and Galileo harm? Would he ever have? There's no reason to think he would. Father, I've been looking over your shoulder ever since he tried to kill us the first time. And that's the first time we know about. And all I could see over your shoulder, all the way from England, was Montepulciano with a raised dagger. Say I'm a poor excuse for a doctor. Disown me if you wish. But I would act no differently. The only solution to the danger to all of you was the eradication of the madman. I'll never think otherwise, and I'd never act otherwise. You can be assured that it's unlikely that another such enemy will arise. I will accept whatever judgment you make, but I ask you to just accept that I did what I considered the right thing to do." He paused and smiled, now not so grimly. "Besides, when you think about it, what we really did was put Montepulciano in the position of having himself killed. His murder was committed on his behalf, as the fulfillment of his own plan. All we did was give him a carriage ride to his destiny."

What could I say? "I have agonized over this very question many times, and every time refused to commission murder. I always attributed such decisions to moral superiority. I've also always had a slight suspicion that it might be moral cowardice. Sometimes it is hard to know what the right thing is, but easy to do it once you make the decision. Sometimes you know what the right thing is immediately, but it is just hard to do. In the case of the eradication of a mortal enemy, it is hard to know what the right thing is and equally hard to do it. Given my age, I should have more wisdom, but I am at least wise enough to admit that I am still not sure about the disposition of Montepulciano. So, what's done is done, and I love you as I always have."

Vito sighed. "So. You must have many questions about how all this transpired."

"To be sure. First, how did you manage this plan? Have you been coming to Italy on a regular basis? How did you organize the underworld conspirators you needed? How.....?"

"First, let me make it a bit more clear by telling you that your underworld source, Amadeo, was the nexus for the whole thing. I know you wanted the world to think I died on the road to Commune de Tirol, and I left intending to keep to your plan. But the more I thought about it, the more worried I became for your safety, so I started correspondence with Amadeo. Remember, he was my source also, the means I had used to watch Federico's back. And that's another way I differed from you. You had always tried to keep Galileo from knowing he had a guardian angel. I'm sure he always knew anyway. I decided to let Federico in on my patronage from the start, and I gambled that if Amadeo could keep my existence a secret, so could Fang. So we had a grand carousel: Amadeo watching Galileo and Fang, me watching you and Fang through Amadeo, Amadeo bringing Guido and Bufana on board, everybody riding merrily along."

"And you did this all by correspondence."

"Using the couriers you had set in place for

correspondence with Mister Levin or Baker or whatever name comes to mind when you think of the English tailor who has done so well by us."

"But Amadeo was just here last night. He acted as if he thought you dead, and that he was only brought into the plot at the last moment by his friend Guido, who in turn was asked to carry out the plot by a mysterious stranger. And in the background was another stranger who looked somewhat like me, and who I assumed was you. That was a bit confusing. If you had come to Italy to organize this thing, why would Amadeo think you dead, and why wasn't he in on it in the first place?"

"He didn't, and he was. Amadeo respects you as a man of honor, and he made a vow that he would never mention anything I was up to concerning Federico to you, nor would I know about your secret dealings concerning Galileo. He kept his vow, for which I'm most grateful, because I didn't want you to know beforehand we were planning the death of Montepulciano. You're argument against such an action is much weaker after the fact."

"I see that. I also suspect that the mysterious stranger in all this was Federico. He was your leader here in Italy."

"He was only too glad to do this. He loves you like a father, for how you accepted him, introduced me to Amadeo so I had someone to act as a watchdog, and how you saw to my safety when things got dangerous here in Italy. He compared you to his own father and saw which one of you to reject."

"That's a great irony. I have had certain resources all these years, people I could trust, people of the streets, and used them to protect those I loved. I send you to off to England, letting the world of Italy, including these resources and even your best friend, think you are dead, and you immediately begin using everyone I just mentioned to protect me. And in the bargain, Galileo. It is a grand opera, but my part in it was written out of the libretto. I had written that part as the man who saved Galileo, never to be recognized as such, but a noble character anyway. In fact, I was just a stage hand, throwing light on all the real characters in the show."

"Oh you saved Galileo many times. And you're a Jew who says he's a Catholic but isn't, and isn't really a Jew either. You were all these things and you've never been any of them. Believe me, you gave me a religion. Yours, which is all of them, and therefore none of them. Ask your friend Galileo what is the better answer to a vexing problem – anything or nothing. I'll bet he answers the latter. If any answer is correct, there is no correct answer. If none is, you just haven't tried hard enough.

Anna had been sitting quietly. She interjected abruptly. "I can't believe you two. You want the same things, use the same people to help, help the same people you love, and yet you don't understand each other. I understand you both. And I love you both. I'm just a bit angry at you, Vito, for doing all this without telling us. And for showing up at the door as if you had never been away. A little angry, although not enough to avoid a smile. But why didn't you just let us know what you were up to?"

"I've already told you. I didn't want to debate it. Besides, it all came together rapidly, as if by providence. I couldn't have told you of a plan I didn't yet have. Some of it rested on Sir Thomas Heedon's mission, which I learned of only a short time ago. Some of it relied on intelligence I received concerning Montepulciano's return to Padua and his searching for scurrilous for assassins. Some of it relied on Federico and his own underworld contacts. Of course, there is Bufana, the start of the whole thing. Guido knew she was in Padua, had served in places like the pastry shop, had a soft spot for Galileo and a hard spot for the Bellarminos. For that matter, I met Guido only three days ago. Here was a down and out man with a game leg, eager to help Il Signore Fishbeingnola. Heedon probably took me on because of my fluency in Italian, without knowing my true background or my real purpose. All of this had to stay between me and Federico and Amadeo and Guido until the deed was done. Only then could I face you with honesty."

Anna had left her chair and was standing behind Vito, rubbing his shoulders. She shook her head and smiled. "I believe you did the right thing. Of course I would say that, being one of the beneficiaries. But the priest was a bad man, and helping him cause his own death was a blessing to his church and his order." Then she had another thought. "But...Bufana Giacometti? How did you enlist her aid? This is all too impossible."

"Yes, it is. I arrived in Venice, Fang had found Guido, Guido found Bufana,, and we conceived an improbable plot to destroy a vile enemy. Just in time. It almost makes me believe in God."

"Ah," said Anna. "There might be hope for your salvation. I've given up on your father, although I can get him into heaven on my say so. I may only get one such favor, so you had better get Catholicism or be a really good Jew."

"How about the Church of England? The vicars at Trinity College seem to think I'm a proper Christian."

"That might work too."

Chapter Twenty-one

Excerpts from the memoirs of Salvatore Isaac Fishbeingnola, 1643

My friend Galileo died in January of last year. He had developed many health problems in those later years, becoming essentially blind, so his confinement to Arceti wasn't so troubling to him. He was just short of his 78th birthday.

Being 78 makes me a very old man, although I just became a grandfather for the first time, which gives me great pleasure. I can't be near the child, however, since his father and my only son remain in England. Vito, to all around him Doctor Victor Fishbourne, is a pillar of his community in the tradition of a true blue-blooded, Anglican Englishman. Just a few people in the whole world know his true background. We managed to keep that a secret while it was a necessity to do so. Now it would be foolish to disturb my son's friends and family by revealing his history.

My darling Anna is still alive, also. Yes, she is an old woman now, but still beautiful. She is really the only mother Vito ever knew. Gioia was with me such a short time, just a year or so, and Anna has been mistress of my house for more than forty years.

I changed my name back to Salvatore Isaac now that the dreaded Montepulciano is gone. There was one other reason for the change.

My grandson was born on Christmas Day of last year, in Woolsthorpe, Lincolnshire, in the Woolsthorpe Manor House. His mother was invited to bear her child there by the mistress of the manor, who was having a child during the same period, and was being attended by Doctor Victor Fishbourne. My son and the wife I've never met chose the name Isaac for their child, after my original middle name, so I could do nothing else but reassume that name. He was special, because his birth brought something to mind. I recalled the discussions Galileo and I had about the fact that he was born the same year Michelangelo died. Did Galileo's God, not wanting to waste a really fine soul, give Michelangelo's soul to Galileo? We both thought such things unlikely, but I couldn't help but note the fact that on the Julian calendar, which England still uses, Isaac Fishbourne was born the year that Galileo died.

There is one more oddity of note. The mistress of Woolsthorpe also bore a son that same day. Her husband, farmer Isaac Newton, died before his son Isaac was born. I'm sure he would have been a proud father, but I wouldn't apply the transfer of Galileo's soul to little Isaac Newton. He will undoubtedly be just another apple farmer.